FV

621.381 C348p
CASTELLUCIS
    PULSE AND LOGIC CIRCUITS
                                    7.95

PULSE AND LOGIC CIRCUITS

PULSE AND LOGIC CIRCUITS

RICHARD L. CASTELLUCIS

VNR VAN NOSTRAND REINHOLD COMPANY

NEW YORK   CINCINNATI   ATLANTA   DALLAS   SAN FRANCISCO
LONDON   TORONTO   MELBOURNE

Electronics Technology Series
Consulting Editor
Richard L Castellucis

Van Nostrand Reinhold Company Regional Office:
New York   Cincinnati   Atlanta   Dallas   San Francisco

Van Nostrand Reinhold Company International Offices:
London   Toronto   Melbourne

Copyright © 1976 by Litton Educational Publishing, Inc.

Library of Congress Catalog Card Number: 76-26954
ISBN: 0-442-21476-6

Manufactured in the United States of America

Published by Van Nostrand Reinhold Company
450 West 33rd Street, New York, N.Y. 10001

Published Simultaneously in Canada by Van Nostrand Reinhold Ltd.

15 14 13 12 11 10 9 8 7 6 5 4 3 2 1

Library of Congress Cataloging in Publication Data

Castellucis, Richard L
   Pulse and logic circuits.

   (Electronics technology series)
   Includes index.
   1. Pulse circuits.   2. Logic circuits.
I. Title.
TK7868.P8C38   621.3815'34   76-26954
ISBN 0-442-21476-6

Foreword

In recent years, there has been a substantial increase in the demand for skilled electronics technicians due to the continued growth of the electronics industry, and the resulting increase in the development of electronic devices, systems, and appliances for industry, research, medicine, and the consumer. The need for technicians to design, build, test, install, service, repair, and replace these electronic products is being met by numerous electronics technology instructional programs. These programs are found in several different settings including the technical high school, adult continuing education, industrial inservice training, and two-year community and junior colleges and technical institutes.

The Electronics Technology Series is designed for use primarily by students at two-year colleges and technical institutes. However, because each text was designed to meet specific instructional goals, it is felt that the texts will be valuable to electronics technology programs in any of the other settings listed previously.

The authors and editors of this series established the following primary goals: readability, accuracy, currency, and student involvement. Each text was written to a controlled reading level suitable for college level students in technical programs. Each text represents the current state of the art in electronics technology as it is commonly practiced in industry. Electronics theory is presented to support the concepts which are to be applied to practical situations. A competence in algebra and trigonometry will meet the mathematics requirements of this series of texts. Finally, each text is designed to involve the reader in a total learning experience. Each text unit lists the competencies that the reader is expected to demonstrate after mastering the concepts given in the unit. Extended Study Topics are presented to challenge the motivated reader to go on and pursue more complex areas of interest. In addition, each text includes Laboratory Exercises to provide the reader with experience in working with actual electronic components in practical circuits.

To verify the technical content, each text in this series was classroom tested over a period of several years with two-year electronics technology students.

Preface

With the introduction of the topics of pulse and logic circuits, the reader departs from the world of dc and sinusoidal waveforms. The reader now must learn how to shape these two waveforms to meet his needs. This text proceeds in a logical manner to show the reader how to change a sine wave to a square wave and then how to convert this waveform to spikes or pulses. The method of stretching and shortening these pulses and the creation of a series of pulses from dc is shown. The content presented includes the binary number system, Boolean algebra, multivibrators, and logic circuits with a logic full adder.

Pulse and Logic Circuits is intended for use at the post-secondary technician instruction level for a first course in waveshaping, pulse, and digital techniques. This study will lay the foundation for a second course in Digital Logic Circuits.

The format used in *Pulse and Logic Circuits* emphasizes two factors: (1) the student technician learns by doing, and (2) all laboratory exercises should be realistic and use up-to-date components and devices. Each content unit is centered around a student learning activity of the "hands-on" type. Where necessary, the reader will be led through the assignment step-by-step. At other times, the reader will be told to use his imagination and previous learning experiences to complete the assignment. Approximately 90 percent of the text deals with the use of integrated circuits.

Each unit contains the learning activity, adequate theory concerning the activity, the importance of this unit to the overall topic of pulse circuits, sample problems, and a progress evaluation section.

SECTION I

This section consists of seven units which lay the foundation for pulse and logic circuit generation. Nonsinusoidal waveshapes and their shaping circuits are covered in detail. Each of these topics is necessary if the student is to have a good understanding of pulse circuits as applied to digital electronics. Each topic is presented so as to lead the student toward an understanding of pulse or square wave generation. Units 1, 2, and 3 introduce the student to the square wave and its behavior in two other types of circuitry. Unit 4 is presented to acquaint the student with a special application of the transistor. Unit 5 shows how a special device, the UJT, is used to transform a dc voltage into a series of spikes or pulses. Units 6 and 7 present a variety of circuits that produce a rectangular waveshape for special applications.

SECTION II

Units 8 and 9 depart from conventional circuitry to introduce two new topics, binary numbers and Boolean algebra. These units present the arithmetic of pulse circuits. The rectangular wave with its distinct voltage levels (as compared to dc and sinusoidal ac) lends itself to this new arithmetic. Voltage levels whose values are sharply defined become logic levels and, in turn, the digits of the binary number system.

Boolean algebra is included to provide the reader with the background necessary to apply what is observed to a complete system in a logical manner.

SECTION III

In units 10 and 11, pulse theory and the one-zero numbers of the binary system are combined. The symbols of logic electronics are presented. The gates which permit the pulses and spikes to pass on code or command are shown.

A system is presented which uses these gates to perform arithmetic operations. In this system, the levels of the rectangular wave are being decoded in a special way. This section lays the groundwork for a succeeding course in Digital Logic Circuits. In addition, Section III uses a different point of view to reexamine all of the circuits and logic expressions covered in Sections I and II.

Contents

SECTION 1 WAVESHAPING

SECTION 1 WAVESHAPING

Unit 1

Generation of a square wave

The square in geometry is a special case of a rectangle; similarly, the square wave in electronics is a special case of the rectangular wave. Figure 1-1 shows a rectangular wave and the terminology used to describe it. The rectangular wave is distinguished by the fact that the durations of times t_1 and t_2 are not equal. If this wave is to be considered a square wave, time t_1 must equal time t_2. Unlike the geometric square, the sides of the square wave need not be equal; that is, the peak value length need not equal the length of time t_1. Figure 1-2 illustrates a square wave.

The circuit of figure 1-3, page 3, is used to produce or generate a rectangular wave (or square wave).

When switch S is depressed, current flows through resistor R and causes a voltage E across R equal to the product of the current and the resistance. Voltage E remains as long as the switch is depressed. When switch S is released, the current flow is interrupted and voltage E drops to zero.

In figure 1-1, time $t_1 + t_2$ is called the *pulse repetition time* (prt). This time is the interval between the start of two repeating pulses. The pulse repetition time may be called the time of one cycle of the waveform. The reciprocal of the pulse repetition time is the frequency of the pulse or the *pulse repetition rate (prr)*.

$$\text{prr} = \frac{1}{\text{prt}} \qquad \text{Eq. 1.1}$$

For example, if a square wave has a prt of 10 μ sec., its pulse repetition rate is:

$$\frac{1}{10 \times 10^{-6}} = (0.1) \times (10^{+6}) = 100 \text{ kHz}$$

The frequency (prr) is always given in hertz (Hz).

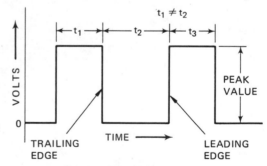

FIG. 1-1   RECTANGULAR WAVE

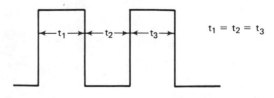

FIG. 1-2   SQUARE WAVE

2

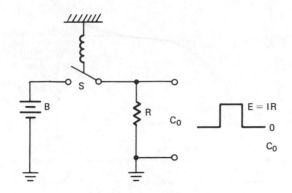

FIG. 1-3 RECTANGULAR WAVE GENERATING CIRCUIT

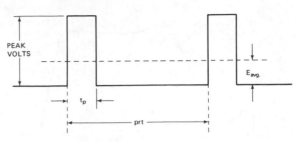

FIG. 1-4 AVERAGE VOLTAGE VALUE
FOR A RECTANGULAR WAVE

The average voltage value for the square or rectangular wave is obtained by dividing the area of the pulse waveform by the frequency (prr). The area of the pulse is equal to the product of pulse width and the peak voltage of the pulse, figure 1-4.

PROBLEM 1

If the peak voltage = 15 volts, t_p (pulse width) = 20 μ sec., and prt = 60 μ sec., solve for (a) prr and (b) E_{avg}.

(a) prr $= \dfrac{1}{prt}$

$= \dfrac{1}{60 \times 10^{-6}} = $ **15 kHz**

(b) $E_{avg.}$ $= \dfrac{(E_p)(t_p)}{prt}$

$= \dfrac{(15)(20 \times 10^{-6})}{60 \times 10^{-6}}$

$= $ **5 volts**

Another quantity which is important to an understanding of square wave circuits is a ratio known as the duty cycle. The duty cycle relates the pulse width (t_p) to the total pulse repetition time (prt) and is given in percent. Eq. 1.2 is used to determine the duty cycle.

duty cycle $= \dfrac{t_p}{prt}$ **x (100) (percent)**

Eq. 1.2

PROBLEM 2

For the values given in Problem 1, find the duty cycle.

duty cycle $= \dfrac{t_p}{prt} = \dfrac{20 \times 10^{-6}}{60 \times 10^{-6}} \times 100$

$= 33\%$

Note that the peak voltage and the average voltage have the same ratio.

The voltage waveforms shown to this point are purely theoretical; that is, the voltage appears to rise from zero to its peak value in zero time. Similarly, the voltage appears to drop from its peak value to zero in zero time. Practically, however, a certain amount of time is required for the voltage to rise to its peak value and to return to zero. The time required for the waveform to rise from 10% of its peak value to 90% of its peak value is called the *rise time*, t_R. A comparison of the rise time, the fall time, and the pulse duration are shown in figure 1-5, page 4.

A typical square wave pulse is shown in figure 1-5, page 4. If E_{pk} is equal to 10 volts, point A will be at a value of 1 volt and point B will be at a value of 9 volts. E_{pk} is the peak value of the pulse regardless of the duration of that value. Therefore, the time required for this pulse to rise from 1 volt to 9 volts is the rise time (t_R). The *fall time* (t_F) is the time during which the pulse decays or falls from point C (90% of E_{pk}) to

3

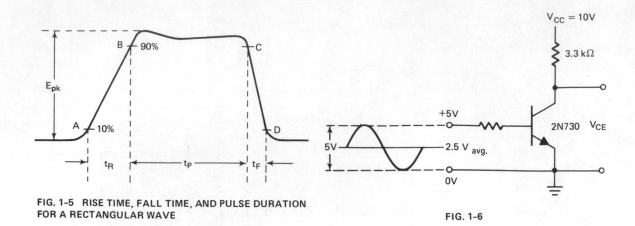

FIG. 1–5 RISE TIME, FALL TIME, AND PULSE DURATION FOR A RECTANGULAR WAVE

FIG. 1–6

point D (10% of E_{pk}). The time or duration that the pulse actually exceeds 90% of E_{pk} is the pulse duration time.

LABORATORY EXERCISE 1-1:   SQUARE WAVE FAMILIARIZATION

PURPOSE

- To acquaint the student with square wave measurement and a method of square wave generation.

MATERIALS

1 Sine-wave generator
1 Power supply, dc, 0-30 volts
1 Transistor, 2N730
1 Resistor, 3.3 k ohm
1 Resistor decade box
1 Oscilloscope

PROCEDURE

A.   Construct the amplifier circuit shown in figure 1-6.

   Note: Since it is the purpose of this exercise to investigate and analyze a square wave, the amplifier has been designated for the student.

B.   1. Apply a sine wave of 5 volts peak to peak as shown in figure 1-6. Use a frequency of 1000 Hz.

   2. Figure 1-7, page 5, shows the family of collector curves for the transistor of the test circuit. The proper load line is drawn on the figure and the base current waveform produced by the sine wave is shown.

   3. Using your knowledge of transistor amplifiers, sketch the voltage waveform you can expect to measure from the collector to ground (since the emitter is at ground potential, this value is actually V_{CE}).

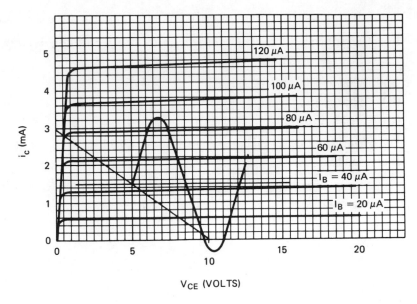

FIG. 1-7 2N730 COLLECTOR CURVES

C. 1. Increase the output of the signal generator to 15 volts peak to peak. This change will cause the base current waveform to extend beyond the limits of the load line. As a result, the transistor will operate in the cutoff regions and in its saturated region for a portion of the input sine wave.

2. Sketch the expected waveform to be measured from the collector to ground (V_{CE}).

D. 1. Connect an oscilloscope between the collector and ground. Does the waveform displayed on the oscilloscope resemble the sketch drawn in step C.2.?

2. The method by which the square wave in this laboratory experience is generated is called the overdriven amplifier method.

E. 1. Measure and record the peak voltage, E_{pk}, of the waveform.

2. Measure and record the rise time, fall time and pulse duration of the generated square wave.

F. Slowly increase the amplitude of the sine wave being applied to the amplifier. State what happens to the rise time, the fall time, and the pulse duration.

STUDENT REVIEW

Questions in this section are designed to evaluate student understanding of the concepts of this unit.

(R-1) Name one disadvantage of generating square waves as demonstrated in this unit.

(R-2) What distinguishes a square wave from a rectangular wave?

(R-3) For the square wave generated in Laboratory Exercise 1-1, answer the following questions:

(a) What is the prt?

(b) What is the prr?

(c) From step E of the procedure, what is $E_{avg.}$?

(R-4) Does the duty cycle change when the amplitude of the input signal is changed?

(R-5) Name one advantage of producing a square wave by the overdriven amplifier method.

Integrators and differentiators

OBJECTIVES

After studying this unit, the student will be able to:

- State the difference between integrators and differentiators
- Calculate time constants
- Discuss the importance of time constants to waveshaping and timing circuits

THE RC TIME CONSTANT

Before beginning the discussion of the resistor-capacitor RC time constant circuit, a brief review of the charging property of capacitors will be helpful. If a capacitor is connected to a battery as shown in figure 2-1, the capacitor will charge (develop a potential across its plates). The amount of charge will equal E_B if SW_1 remains closed for a long enough period of time.

When SW_1 is closed, current I_C will begin to flow at some initial rate. The capacitor will begin to develop a charge on its plates at this same rate. As the capacitor develops its charge, it will appear to be another voltage source in series with and opposing E_B, figure 2-2.

As a result, a current I_O will attempt to flow. Current I_O will alter the rate of current flowing to the capacitor. This change in the

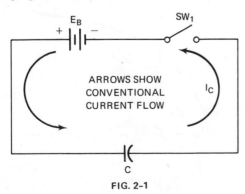

FIG. 2-1

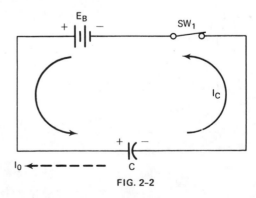

FIG. 2-2

current rate of flow will change the rate at which the capacitor is developing its charge.

If a resistor is placed in the circuit, figure 2-3, the initial current flow in the circuit can be calculated since the only opposition initially to this charging current (I_C) is the resistor.

Once the resistor is inserted in the circuit and the switch is closed, the initial current in the circuit, I_C, equals E_B/R. At the instant the switch is closed, the capacitor has a zero charge and all of E_B must be dropped across R.

As the capacitor begins to charge, current I_O develops and the initial current, I_C, begins to decrease, figure 2-4. As the capacitor charge increases, current I_C continues to decrease. When the charge on the capacitor equals the battery potential, then zero current flows in the circuit. As shown in figure 2-4 this decline in current is not linear but shows an exponential decay.

Ohm's Law ($E_R = I_R R$) indicates that the voltage drop across R must be decreasing since the current in the circuit is decreasing. In other words, the voltage across a resistor is equal to the current through the resistor multiplied by the value of the resistor. The exponential decrease in current means that the voltage across the resistor also decreases exponentially, figure 2-5.

Kirchhoff's voltage law states that the voltage drops around a closed loop (circuit) must equal zero. This means that at any time during the charge cycle of the capacitor, the following statement can be made.

$$(E_B) - (E_R) - (E_C) = 0 \qquad \text{Eq. 2.1}$$

The parentheses are used to designate absolute values. Initially, E_R equals E_B and E_C is equal to zero. To maintain the balance in Eq. 2.1 as E_R decreases exponentially, E_C must increase exponentially, figure 2-6.

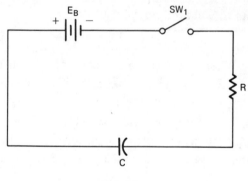

FIG. 2-3

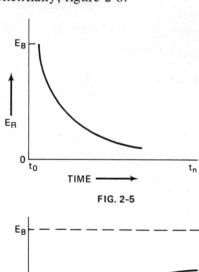

FIG. 2-5

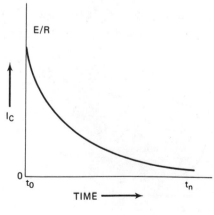

FIG. 2-4

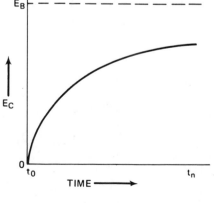

FIG. 2-6

All quantities that grow or decay exponentially do so in the following manner. In the first time span, they will grow or decay to approximately 63% of their maximum potential. For example, if the battery potential in figure 2-3 is 100 volts, then in the first time span of charge for the capacitor it charges to approximately 63 volts. In two time spans, the charge reaches approximately 86% of the maximum value or 86 volts.

In the RC charge circuit of figure 2-3, the length of one time constant (one time span) is equal to the product of the resistor value and the capacitor value. The time constant is given in seconds.

$$\text{TC (time constant)} = RC \qquad \text{Eq. 2.2}$$

PROBLEM 1

If the resistor in figure 2-3 has a value of 1 kΩ and the capacitor has a value of 1μF, calculate the length of one time constant for the circuit.

$1\,k\Omega \;\; = 1 \times 10^3$ ohms
$1\,\mu F \;\; = 1 \times 10^{-6}$ F

Time (of one time constant) = TC, in seconds

$TC \;\; = R \times C$
$TC \;\; = 1 \times 10^3 \times 1 \times 10^{-6}$
$\quad\;\;\; = 1 \times 10^{-3}$ seconds = **1 millisecond**

If the time of one time span for charge is called one time constant, then the time of two time spans of charge is called two time constants, and so on.

Figure 2-7 shows the Universal Time Constant Graph for RC time constant circuits. Note that the graph includes only five time constants. For all practical purposes, a capacitor reaches its maximum charge within five time constants. Curve A is the charge curve and curve B is the discharge curve.

PROBLEM 2

For the circuit of figure 2-8, calculate the following:

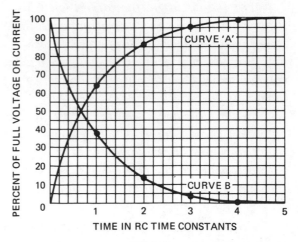

FIG. 2-7

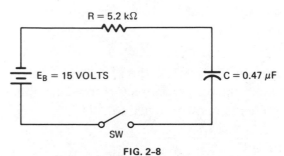

FIG. 2-8

(a) the RC time constant
(b) the voltage on the capacitor at the end of one time constant
(c) the initial current I_C that will flow in the circuit

(a) The RC time constant is equal to (R) x (C)

$TC \;\; = (R) \times (C)$
$TC \;\; = (5.2 \times 10^3) \times (0.47 \times 10^{-6})$
$TC \;\; = 2.44 \times 10^{-3}$ seconds

(b) At the end of one time constant, the voltage across the capacitor is equal to 63% of the maximum value.

$E_C \;\; = 0.63 \times E_B$
$E_C \;\; = 0.63 \times 15$
$E_C \;\; = 9.45$ volts

(c) Current (I_C) is equal to E_B divided by R.

$I_C \;\; = E_B/R$
$I_C \;\; = 15/5.2 \times 10^3$
$I_C \;\; = 2.88 \times 10^{-3}$ amperes

For the circuit in figure 2-8, what is the voltage across the capacitor at the end of two time constants? (R2-1) (See figure 2-8, page 9.)

For the circuit in figure 2-8, what is the voltage across the capacitor at the end of 2.6 time constants? (R2-2)

(Note: to solve question (R2-2), the student should refer to the Universal Time Constant Graph, locate 2.6 TC, follow this value to the point that it reaches the curve, and multiply the value on the curve by 15 volts.)

To determine the voltage of the capacitor at some increment of time not equal to a time constant value, the following equation can be used.

$$E_C = (E_B) \times (1 - \epsilon^{-t/RC}) \qquad \text{Eq. 2.3}$$

In Eq. 2.3, t is the increment of time at which the voltage across C is to be determined and ϵ is a constant equal to 2.718.

PROBLEM 3

For the circuit shown in figure 2-9, find the voltage across C, 3 milliseconds after the switch is closed.

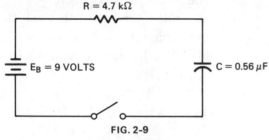

FIG. 2-9

$$E_C = E_B \times (1 - \epsilon^{-t/RC})$$
$$E_C = 9 \times (1 - \epsilon^{(-3 \times 10^{-3})/(2.63 \times 10^{-3})})$$
$$E_C = 9 \times (1 - \epsilon^{-1.14})$$
$$E_C = 9 \times (1 - 0.319)$$
$$E_C = 9 \times (0.681)$$
$$E_C = \textbf{6.13 volts}$$

Eq. 2.3 can be used to solve for the voltage across the capacitor at any increment of time. This equation provides greater accuracy than can be obtained using the Universal Time

Constant Graph. In Problem 3, the time for one time constant is 2.63×10^{-3} second and the time of charge is 3×10^{-3} second. This means that the capacitor must be charged for approximately 1.14 time constants. The use of these numbers in the Universal Time Constant Graph would be very difficult and would give rise to large errors.

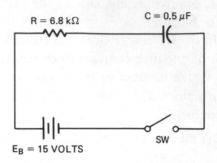

FIG. 2-10

In figure 2-10, what is the voltage across C after 0.5 milliseconds? (R2-3)

For figure 2-10, how long will it take the capacitor to charge to 6 volts? (R2-4)

Note: The solution to question (R2-4) differs slightly from previous solutions. The steps to be followed in this solution are:

1. *Write the charge equation.*
    $$E_C = E_B \times (1 - \epsilon^{-t/RC})$$

2. *Substitute values in this equation.*
    $$6 = 15 \times (1 - \epsilon^{-t/RC})$$

3. *Substitute x for t/RC*
    $$6 = 15(1 - \epsilon^{-x})$$

4. *Solve for x*
    $$\frac{6}{15} = (1 - \epsilon^{-x})$$
    $$0.4 = 1 - \epsilon^{-x}$$
    $$0.4 - 1 = \epsilon^{-x}$$
    $$-0.6 = -\epsilon^{-x}$$
    $$0.6 = \epsilon^{-x}$$
    $$x = 0.51$$

5. Substitute value of x in t/RC to find t.

$$x = t/RC$$
$$0.51 = t/3.4 \times 10^{-4}$$
$$(0.51) \times (3.4 \times 10^{-4}) = t$$
$$t = 1.73 \times 10^{-4} \text{ second}$$

Use figure 2-11 to solve for the following quantities.

(a) Find the RC time constant (R2-5)
(b) Find the charge on C in the two time constants (R2-6)
(c) Find the charge on C in 3×10^{-3} second (R2-7)

Figure 2-12 shows the circuit of figure 2-11 with the addition of a two-position switch. If the switch is placed in position A, the battery will be in the circuit and the capacitor will attempt to charge to the battery potential. With the switch in position B, the capacitor will attempt to discharge. The time constants for charging and discharging will be the same. As stated previously, the product R x C is the time constant. The Greek letter τ (tau) is used to represent R x C. Therefore, for figure 2-12,

$$\tau = R \times C$$
$$\tau = 4.7 \times 10^{-3} \text{ second}$$

If the switch is placed in position A and is held there for 4.7×10^{-3} second, the capacitor will charge to 63% of E_B or 9.45 volts. At the end of this time, if the switch is placed in position B, the capacitor will begin to discharge. If the switch is held in position B for 4.7×10^{-3} second, the capacitor will discharge to approximately 37% of its charged value at the start of the discharge cycle. Thus, the capacitor will discharge to 37% of 9.45 volts or 3.5 volts.

PROBLEM 4

Using figure 2-12, find the voltage of C when the switch is first placed in position A for 3×10^{-3} seconds, and then when the switch

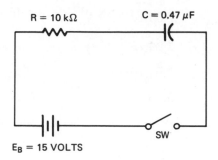

R = 10 kΩ   C = 0.47 μF

E_B = 15 VOLTS

FIG. 2-11

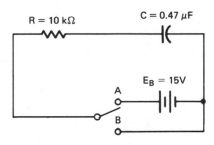

R = 10 kΩ   C = 0.47 μF

E_B = 15V

FIG. 2-12

is placed in position B for 3×10^{-3} seconds. Assume that the initial charge is zero.

1. Write the charge equation.

$$E_C = E_B (1 - \epsilon^{-t/RC}) \text{ or } E_B (1 - \epsilon^{-t/\tau})$$

$$E_C = 15 \cdot (1 - \epsilon^{(-3 \times 10^{-3})/(4.7 \times 10^{-3})})$$

$$E_C = 15 (1 - \epsilon^{-0.638})$$

$$E_C = 15 (1 - 0.528)$$

$$E_C = 15 (0.471)$$

$$E_C = \textbf{7.07 volts}$$

E_C = 7.07 volts is the voltage on the capacitor when it begins its discharge cycle (when the switch is placed in position B). The voltage on the capacitor after the discharge cycle is given by Eq. 2.4.

$$E_{cf} = E_{ci} (\epsilon^{-t/\tau}) \qquad \text{Eq. 2.4}$$

where
E_{cf} = final capacitor voltage after discharge

E_{ci} = initial capacitor voltage before discharge

$$E_{cf} = 7.07 \, (\epsilon^{\,(-3 \times 10^{-3})/(4.7 \times 10^{-3})}$$

$$= 7.07 \, (\epsilon^{-0.638})$$

$$= 7.07 \, (0.528)$$

$$E_{cf} = 3.73 \text{ volts}$$

Refer to figure 2-12 and solve for the following quantities.

(a) Find the voltage on C, if the switch is in position A for 6 milliseconds and then is placed in position B for 4 milliseconds. (R2-8)

(b) Find the voltage on C if the switch is in position A for 6 milliseconds, is placed in position B for 4 milliseconds, and is returned to position A for 3 milliseconds. (R2-9)

> *Note: To solve part (b) of Problem 4, part (a) must be solved first. Designate the voltage on the capacitor at this time as e_C. When the switch returns to position A for 3 milliseconds, the capacitor continues to charge toward E_B. However, the charge value is only $E_B - e_C$ since part of E_B or e_C is already across the capacitor.*

The recharging of the capacitor is given by Eq. 2.5.

$$E_C = E_B - e_C \, (1 - \epsilon^{-t/\tau}) \qquad \text{Eq. 2.5}$$

THE INTEGRATOR

If the battery and switch of figure 2-12 are replaced with a pulse generator, figure 2-13, and the pulses are controlled as shown in figure 2-14, then the result is the automatic switching of the circuit in figure 2-12.

If the square wave output from the pulse generator, figure 2-15a, is applied to the circuit shown in figure 2-13, then the waveshapes are as shown in figure 2-15b.

Note that the waveshapes in figure 2-15 differ for the resistor and the capacitor. The voltage across the capacitor charges expon-

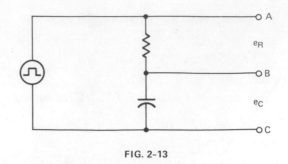

FIG. 2-13

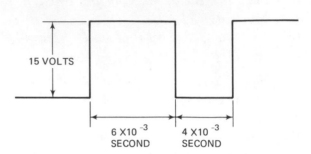

15 VOLTS

6 X10^{-3} SECOND   4 X10^{-3} SECOND

FIG. 2-14

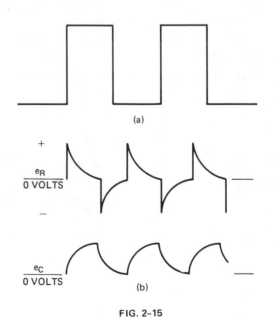

(a)

e_R
0 VOLTS

e_C
0 VOLTS
(b)

FIG. 2-15

entially and will continue to charge as long as the pulse is present. This charging pattern is called *integrating action*. If the pulse duration is too short, that is, if it does not give the capacitor sufficient time to charge fully, then the output waveshape across the capacitor will be greatly distorted.

LABORATORY EXERCISE 2-1:   ANALYSIS AND MEASUREMENT OF AN RC INTEGRATOR CIRCUIT

PURPOSE

- To acquaint the student with the techniques of analyzing an RC integrator.

MATERIALS

1 Square wave pulse generator

1 Oscilloscope

1 Resistor, 2.2 k ohm

1 Capacitor, 1.0 μ F

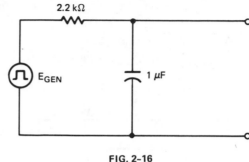

FIG. 2-16

PROCEDURE

A.   Construct the circuit of figure 2-16.

B.   Calculate and record the value of τ = RC.

C.   Apply a 1000-hertz signal of 10 volts peak.

1. What is the prt of a 1000-hertz signal?
2. What is the pulse time (t_p) of this signal?

D.   Calculate the charge on C (E_c peak) during the on time of the pulse.

E.   Using the oscilloscope, measure and draw the output waveform across the capacitor.

F.   Perform the following steps to determine how variations in the prr (pulse repetition rate) affect the waveforms.

1. Decrease the frequency of the square wave to 500 hertz and draw the output waveshape.
2. Increase the frequency of the square wave to 2000 hertz and draw the output waveshape.
3. Increase the frequency to 5000 hertz and draw the output waveshape.

Notice that as the frequency of the square wave input increases, the output waveshape as taken across the capacitor becomes more and more distorted.

SUMMARY

It is apparent that if the waveshape across the capacitor is to closely resemble the input waveshape, then the following relationship must be true.

$$t_p \; \geqq \; 10\tau$$

When this condition is observed, the resulting circuit is known as a *short time constant circuit.*

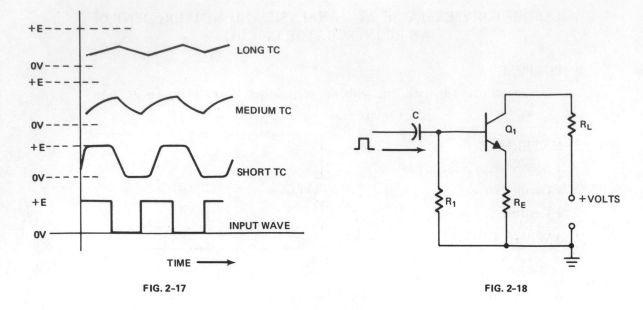

FIG. 2–17   FIG. 2–18

However, the following condition may be true.

$$\tau \cong 10\, t_p$$

For this condition, the resulting circuit is a *long time constant circuit.* When τ (tau) = 10 t_p, a large distortion in the output waveshape is produced.

Circuits for which the relationship between τ and t_p is between the values shown, are known as *medium time constant circuits.*

The effect of the time constant in the output waveform is shown in figure 2-17.

THE DIFFERENTIATOR

The RC differentiator circuit is identical in appearance to the RC integrator circuit. The major difference between the two circuits involves the point where the output waveshapes are measured. For the differentiator circuit, the output is measured across the resistor.

LABORATORY EXERCISE 2-2:   ANALYSIS AND MEASUREMENT OF AN RC DIFFERENTIATOR CIRCUIT

PURPOSE

• To acquaint the student with the RC differentiator circuit

MATERIALS

1 Square wave pulse generator
1 Oscilloscope
1 Resistor, 2.2 k ohm
1 Capacitor, 1.0 μ F

PROCEDURE

Use the circuit of figure 2-16 and follow the procedure steps given in Laboratory Exercise 2-1. For the present exercise, however, the output measurements are to be made across the resistor.

If a short spike or pulse is to designate the start of a positive going pulse, then that pulse is inserted into an RC time constant circuit and the output is measured across the resistor. Measurements taken in this laboratory exercise should indicate that the voltage waveform across the resistor has both positive and negative portions due to the fact that the current during the charge cycle passes through the resistor in one direction and reverses itself during the discharge cycle.

Using the circuit of figure 2-18, page 14, determine the answers to the following questions.

(a) *What relationship between t_p and τ will result in the proper coupling of the pulse to the base of Q_1? (R2-10)*

(b) *What factors must be considered when calculating τ? (R2-11)*

EXTENDED STUDY TOPICS

1. What changes must be made in the charge and discharge equations if the input pulses to an RC integrator are nonsymmetrical?

2. What are the effects on the output waveshape of an RC integrator with a nonsymmetrical input pulse?

Clippers and clampers

OBJECTIVES

After studying this unit, the student will be able to:

- Describe the differences between clipper and clamper circuits
- Demonstrate an understanding of series and shunt clipping through circuit construction
- Construct biased and unbiased clipper and clamper circuits

DIODE CLIPPING

In this unit, the student will investigate the desirable and the undesirable effects of a clipping circuit on a particular waveshape.

Clipping implies that a portion of the waveform is cut off (clipped). Another name for a clipper circuit is a *limiter* circuit. There actually is a distinction between the actions of a clipping circuit and a limiter circuit. The limiter circuit limits the amplitude of the waveshape, figure 3-1. The clipper circuit, on the other hand, cuts off or clips the signal, figure 3-2, page 17. The student should study figures 3-1 and 3-2 to note the differences in the actions of these two circuits. The limiter circuit of figure 3-1 limits the maximum value of the input signal which is allowed to pass. Therefore, if the peak value of the input is seven volts and the limiter circuit is set to limit at 4 volts, then that portion of the sine wave above 4 volts will not pass through the limiter circuit. In other words,

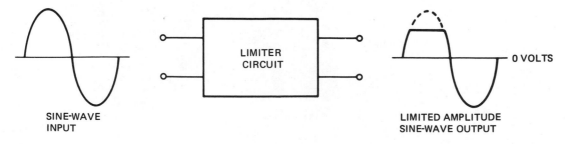

FIG. 3-1   THE LIMITER CIRCUIT

16

a limitation is placed on the peak value of the input sine wave that is allowed to pass.

The clipper circuit, on the other hand, clips off part of the input waveform. In general, the clipper circuit is used to cut off the positive or negative portion of the input waveform. In this text, the general phrase clipping circuit will be used to refer to both of these circuits.

SERIES DIODE CLIPPER

The simplest diode clipper circuit is the *series diode clipper,* figure 3-3. The input and output waveshapes for this circuit are also shown. As the input wave goes positive, the diode is forward biased and current flows through the resistor and diode. An ideal diode is assumed; that is, there is no voltage drop across the diode. Therefore, all of the

voltage will be developed across the resistor. As the input wave goes negative, the diode is reverse biased. In this case (again assuming an ideal diode), there is zero current flow in the circuit. As a result, zero voltage is developed across the load resistor. The output waveform in figure 3-3 shows that the negative portion of the input sine wave has been clipped. The action of the circuit results in the shaping or reshaping of a waveform or signal voltage. If the diode is reversed, figure 3-4, the circuit clips the positive portion of the input waveform.

SHUNT DIODE CLIPPER

To change a series diode clipper to a *shunt diode clipper,* the output is taken across the diode, figure 3-5, page 18. The circuit arrangements for both positive and negative shunt clipping are shown in the figure.

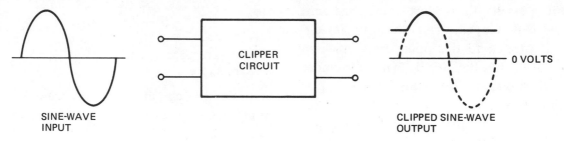

FIG. 3-2   THE CLIPPER CIRCUIT

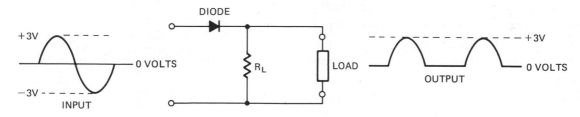

FIG. 3-3   SERIES DIODE CLIPPER CIRCUIT AND WAVESHAPES FOR NEGATIVE CLIPPING

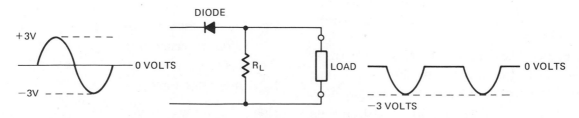

FIG. 3-4   SERIES DIODE CLIPPER CIRCUIT AND WAVESHAPES FOR POSITIVE CLIPPING

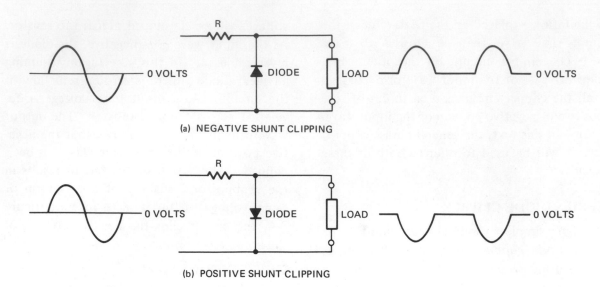

(a) NEGATIVE SHUNT CLIPPING

(b) POSITIVE SHUNT CLIPPING

FIG. 3-5 SHUNT DIODE CLIPPER

When the diode conducts in this circuit (when it is forward biased by the input voltage), the circuit is essentially a short circuit and zero voltage is developed. The diode clipping circuits illustrated to this point are considered to be ideal. In other words, when the diodes are reverse biased (nonconducting), they permit no current flow and have infinite resistance. Similarly, when the diodes are forward biased (conducting) they are assumed to be perfect short circuits and thus develop zero voltage drop. In practical applications, however, diodes are not perfect and do not perform as in the ideal assumption. Diodes do exhibit some resistance both in the forward biased and the reverse biased modes of operation. When the diode is operating in the forward direction, its resistance is usually very low, around 100 ohms. When the diode is operating in the reverse direction, its resistance is about 10 megohms. In other words, the proper operation of the clipper circuit requires that the load resistance value must be such that it does not greatly limit the current flow during forward bias operation, and does not develop too large a voltage drop during reverse bias operation. Mathematically, this statement is expressed as follows:

$$100\,\Omega << R_L << 10\,M\,\Omega$$

PROBLEM 1

For the circuit shown in figure 3-6, assume a diode with the following characteristics:

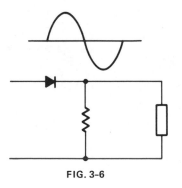

FIG. 3-6

R_f (diode forward resistance) $\approx 100\,\Omega$
R_r (diode reverse resistance) $\approx 100\,M\Omega$
R_L (load resistor) $= 1\,k\Omega$
E_{in} (input sine-wave peak voltage) $= 3$ volts

Draw the output waveform and calculate the voltage levels of this waveform.

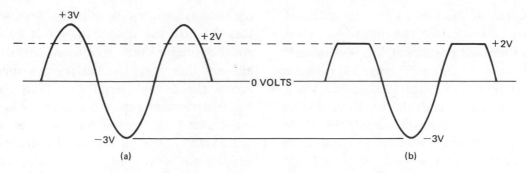

FIG. 3–7   INPUT AND OUTPUT OF NONZERO CLIPPER CIRCUIT

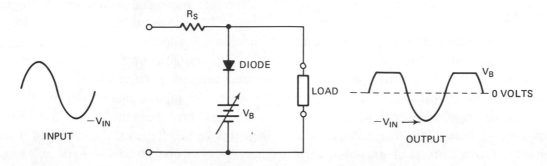

FIG. 3–8   THE BIASED SHUNT CLIPPER

Positive Half Cycle:

During the positive half cycle, the diode is conducting. Current flow in the circuit is given by Ohm's Law, $I = E/R$.

$$I = \frac{3}{R_f + R_L} = \frac{3}{100 + 1000} \text{ ohms}$$

$$I = \frac{3}{1100} \text{ ohms}$$

$$I = 0.00272 = 2.72 \text{ mA}$$

At this time, the voltage drop across the load resistor is:

$$E_{R_L} = I \times R_L$$
$$= (2.72 \times 10^{-3}) \times (1.0 \times 10^{3})$$
$$= 2.72 \text{ volts}$$

This means that 0.28 volt is being dropped across the diode. This value may be a large enough signal loss to affect the operation of a transistor amplifier or a transistor switch.

Negative Half Cycle:

During the negative half cycle of the input signal, the voltage divider represented by R_r and R_L is great enough to drop only 30×10^{-6} volt across R_L.

Problem 1 shows that the size of R_L is increased to insure a greater voltage drop across it while the diode is forward biased. Additionally, there is a larger voltage drop across R_L when the diode is reverse biased. To find the optimum value of R_L for the series diode clipper circuit (or for the shunt diode clipper), use the following equation:

$$R_L = \sqrt{R_f \, R_r} \qquad\qquad \textbf{Eq. 3.1}$$

where R_f = diode forward resistance

R_r = diode reverse resistance

For Problem 1, calculate the optimum value of R_L. (R3-1)

Using the value of R_L found in (R3-1), calculate the voltage drop across R_L while the diode is forward biased. (R3-2)

What is the difference between shunt diode clipping and series diode clipping? (R3-3)

BIASED DIODE CLIPPERS

The clipping circuits examined to this point clip or remove either the positive or negative half of the sine wave. If it is desired to remove a section of the input sine wave, such as the small portion of the positive half cycle as shown in figure 3-7, page 19, then the circuit of figure 3-8, page 19, is used. When the input signal reaches the level required so that the diode is forward biased, the diode will conduct (short out) and produce a dc level at the output terminals which is equal to the bias battery value.

PROBLEM 2

If the input signal of figure 3-8 is 4 volts peak (8 volts peak–to–peak), and one volt is to be clipped from the positive peak, then V_B is adjusted for 3 volts. As the input signal starts in the positive direction, the diode remains reverse biased (having a very high resistance). For all practical purposes, zero current flows in the circuit. As soon as the input signal reaches the potential necessary to forward bias the diode, the diode conducts and becomes a short circuit. At this time, it appears that only the battery is connected across the output terminals. Thus, all of V_B is developed across the load. When the input signal falls below the value at which the diode is forward biased, the diode again appears to be an infinite resistance and all of the signal current flows through the load to recreate the signal waveshape at this time. Figure 3-9 shows these two cases for the biased diode clipper.

The variable battery is convenient in a circuit when it is desired to change the point at which the input signal is clipped. For most situations, however, the point at which the signal is to be clipped is constant and it is not necessary to use a variable battery. For example, note in figure 3-10, page 21, that a

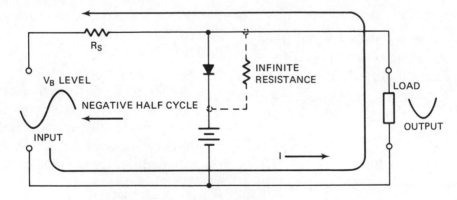

(a) EQUIVALENT CIRCUIT WHEN DIODE IS REVERSE BIASED

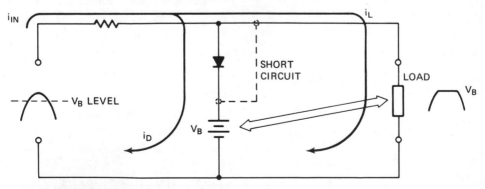

(b) EQUIVALENT CIRCUIT WHEN DIODE IS FORWARD BIASED (INPUT EXCEEDS V_B)
($i_{IN} = i_D + i_L$)

FIG. 3-9

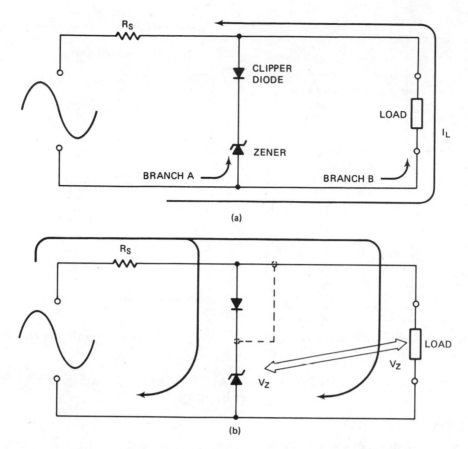

R_S

CLIPPER
DIODE

LOAD

I_L

ZENER

BRANCH A

BRANCH B

(a)

R_S

LOAD

V_Z

V_Z

(b)

FIG. 3-10   USING A ZENER DIODE TO BIAS THE SHUNT DIODE CLIPPER

zener diode is used in place of the variable battery in figure 3-9. (A small fixed battery can also be substituted for the variable battery.) The operation of the circuit in figure 3-10 differs slightly from that of the battery circuit. During the negative half cycle of the input sine wave, the zener diode is forward biased. The clipper diode, on the other hand, is reverse biased; thus, branch A does not allow any current flow. All of the signal current must go through branch B. As a result, the entire signal voltage is developed across the load and the output voltage equals the input voltage.

What is the relationship of the load resistance to R_S? (R3-4)

As the input signal goes positive, for the circuit in figure 3-10, the clipper diode cannot be forward biased until the input signal is more positive than the zener voltage. Even though the zener diode is reverse biased at this point, it does conduct since a zener can conduct in the reverse direction. As the zener conducts, a constant voltage is developed across its terminals. This voltage will remain constant for varying amounts of current flow through the zener diode. The input signal diode must overcome this voltage drop across the zener diode before the clipper diode will conduct. When the clipper diode conducts, it becomes a virtual short circuit and the zener voltage is developed at the load, figure 3-10b.

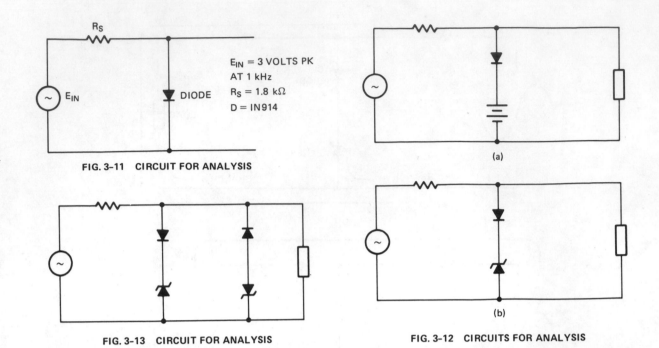

FIG. 3–11   CIRCUIT FOR ANALYSIS

E_{IN} = 3 VOLTS PK
AT 1 kHz
R_S = 1.8 kΩ
D = 1N914

(a)

FIG. 3–13   CIRCUIT FOR ANALYSIS

(b)

FIG. 3–12   CIRCUITS FOR ANALYSIS

LABORATORY EXERCISE 3-1:   ANALYSIS OF VARIOUS TYPES OF DIODE CLIPPERS

PURPOSE

- To analyze the diode clipper circuits of figures 3-11, 3-12, and 3-13.

MATERIALS

1 Sine-wave generator, low impedance output at an output frequency of 1 kHz

2 Diodes, 1N914, or equivalent

2 Zener diodes, 1N4728, or equivalent

1 Resistor, 1.8 kΩ

1 Oscilloscope

PROCEDURE

A. Draw the output waveshapes expected for each circuit in figures 3-11, 3-12, and 3-13.

B. Correct any errors in the resistor value, if necessary.

C. 1. Construct the circuits and measure the signals.

2. Compare the output waveforms with those predicted in step A.

DIODE CLAMPING CIRCUITS

A commonly used simple electronic waveshaping circuit is the *diode clamper.* The function of the diode clamper circuit is similar to that of the clipper circuit with one exception: the clipper circuit distorts or changes the appearance of the input wave while the output wave of the clamper circuit

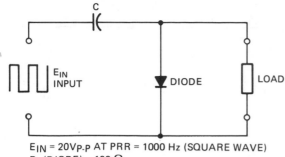

E_{IN} = 20Vp-p AT PRR = 1000 Hz (SQUARE WAVE)
R_f (DIODE) = 100 Ω
R_r (DIODE) = 10 MΩ
R_{GEN}(GENERATOR) = 50 Ω

FIG. 3–19

To check that the proper value of the capacitance is selected, recall that while the diode is reversed biased, C will try to discharge. Since the reverse resistance of the diode is 10MΩ, the time constant τ is:

$$\tau = R \times C$$
$$= 10 \times 10^6 \times 0.66 \times 10^{-6}$$
$$= 6.6 \text{ sec.}$$

Since the diode is reverse biased for only 500 μ seconds, the capacitor does not have enough time to discharge. If the value of the load resistor is only 1 kΩ, then the discharge time constant is equal to:

$$\tau = R \times C$$
$$= 1 \times 10^3 \times 0.66 \times 10^{-6}$$
$$= 0.66 \times 10^{-3} \text{ sec.}$$

Since the time constant is larger than the pulse width of 500 μ seconds, the diode is reverse biased.

The clamping circuit investigated in Problem 3 clamps the output voltage at zero so that the entire signal goes negative from that value. This type of circuit is called a *negative voltage clamper.*

To construct a positive voltage clamper circuit where the signal is clamped at zero but goes positive from that point, what change must be made to the negative voltage clamper circuit in figure 3-19? (R3-5)

The positive and negative clamping circuits shown will clamp the input signal to zero. These circuits are particularly useful when the output of a pulse generator centers about zero and the required signal must be positive or negative going from zero with no distortion.

If the input voltage fluctuates in a clamping circuit, the output does not clamp to zero because the capacitor must be charged to the peak of the input signal. If the input level decreases, and the time constant of the capacitor does not permit it to discharge to the new level, then the output level will shift. This statement can be verified by studying figure

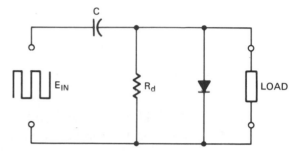

FIG. 3–20 R_d PARALLELED TO IMPROVE CLAMPING

3-18. To compensate for any input voltage fluctuations, a resistor may be placed in parallel with the diode, figure 3-20.

To achieve the proper operation of the circuit, the added resistor, (R_d), must be greater in value than the forward resistance of the diode, (R_f). Additionally, the value of R_d must be less than the reverse resistance, (R_r), of the diode. The value for R_d is determined by the use of the geometric mean equation.

$$R_d = \sqrt{R_f \ R_r} \qquad \text{Eq. 3.2}$$

For the values of Problem 3, R_d is equal to:

$$R_d = \sqrt{(100) \ (10 \times 10^6)}$$
$$R_d = 31.62 \text{ k}\Omega$$

The use of R_d in the circuit means that a change will occur in the time of the discharge time constant. Since R_d is in parallel with the diode (and thus R_r), τ is equal to:

$$\tau = (R_{GEN} + \frac{R_d \ R_r}{R_d + R_r}) \times (C)$$

For the values of Problem 3:

$$\tau = \left[50 + \frac{(31.6 \times 10^3)(10 \times 10^6)}{31.6 \times 10^3 + 10 \times 10^6}\right] \times (C)$$

$$\tau = (50 + 31.50 \times 10^3) \times (C)$$

$$\tau = (31.55 \times 10^3) \times C$$

Since τ is the discharge time constant, it should be approximately 100 times greater than the time during which the diode is reverse biased.

$$\tau = 100 \, t_p$$
$$= (100)(500 \times 10^{-6} \text{ sec.})$$
$$= 50 \times 10^{-3} \text{ sec.}$$

Therefore,

$$C = \tau/R$$
$$= \frac{50 \times 10^{-3}}{31.6 \times 10^{-3}}$$

$$C \approx 1.58 \, \mu F$$

The student should recall from Unit 2 that when R is in ohms, C is in microfarads (μF).

For the circuit in figure 3-21, solve for the following quantities.

(a) R_d (R3-6)

(b) *Value of C*   (R3-7)

(c) τ *on discharge*   (R3-8)

NONZERO CLAMPING

The clamping circuits investigated to this point in the unit are concerned with clamping at a zero reference level only. If the incoming signal is to be clamped at some value other than zero, then the circuit shown in figure 3-22 can be used. For this circuit, the output is clamped at a value of +5 volts.

Does the circuit in figure 3-22 demonstrate positive or negative clamping? (R3-9)

Operation of a Nonzero Clamping Circuit

Figure 3-22a and figure 3-22b illustrate the operation of this type of circuit. At the time the input reaches –10 volts, it and the battery are series additive and the diode is forward biased. The capacitor can charge to the

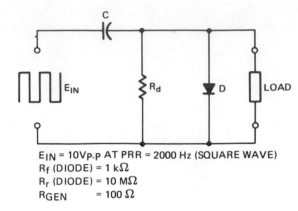

E_{IN} = 10Vp-p AT PRR = 2000 Hz (SQUARE WAVE)
R_f (DIODE) = 1 kΩ
R_r (DIODE) = 10 MΩ
R_{GEN} = 100 Ω

FIG. 3–21 CIRCUIT FOR ANALYSIS

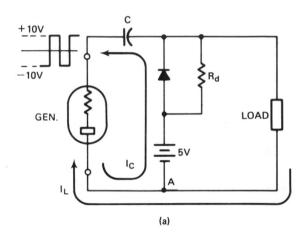

(a)

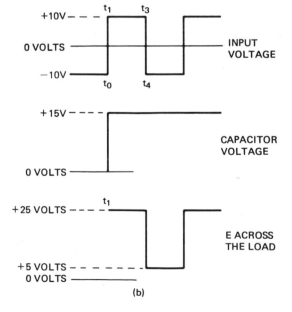

(b)

FIG. 3-22 CLAMPER CIRCUIT WITH WAVEFORMS

sum of the battery voltage and the input voltage. Thus, at time t_o, the capacitor charges to +15 volts with reference to point A. At time t_1, the input voltage reaches +10 volts. The diode is reverse biased and the load sees an equivalent 10-volt generator battery voltage in series with and aiding the capacitor to form a path for the current flow (outer loop). At the same time, a voltage drop of +25 volts occurs across the load. The capacitor voltage and the generator voltage are now added according to Kirchhoff's law. At time t_2, the input drops back to –10 volts in series with

and opposing the capacitor. As a result, the output appears to be +5 volts. Note that the output wave is still 20 volts pk–to–pk; however, its reference voltage is shifted. Another name for nonzero clamping is *bias clamping* or *bias-level clamping*.

Can a zener diode be used to shift the reference voltage in place of a battery? (R3-10)

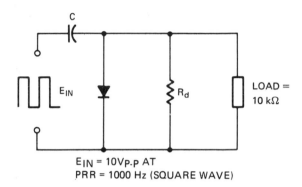

E_{IN} = 10Vp-p AT
PRR = 1000 Hz (SQUARE WAVE)

FIG. 3–23   CIRCUIT FOR ANALYSIS

LABORATORY EXERCISE 3-2: ANALYSIS AND MEASUREMENT OF CLAMPER CIRCUITS

PURPOSE

• To acquaint the student with the function and operation of various clamper circuits.

MATERIALS

1 Square-wave generator, with input at $10V_{p-p}$ at 1000 Hz
1 Diode, 1N914, or equivalent
1 Zener diode, 1N4728, or equivalent
1 Battery, 3V

PROCEDURE

A. Construct and analyze the circuits of figures 3-23, 3-24, and 3-25.

B. Draw the expected waveshapes for the three circuits.

C. Calculate the values of R_d and C required for the proper operation of each circuit. Check values for R_{GEN}, R_f, and R_r.

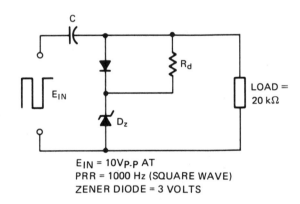

E_{IN} = 10Vp-p AT
PRR = 1000 Hz (SQUARE WAVE)
ZENER DIODE = 3 VOLTS

FIG. 3–24   CIRCUIT FOR ANALYSIS

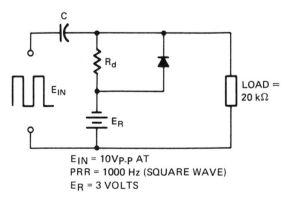

E_{IN} = 10Vp-p AT
PRR = 1000 Hz (SQUARE WAVE)
E_R = 3 VOLTS

FIG. 3–25   CIRCUIT FOR ANALYSIS

27

Name the basic differences between a clipper circuit and a clamper circuit. (R3-11)

For a clamping circuit, what happens if the generator (or source) resistance is too high? (R3-12)

EXTENDED STUDY TOPICS

1. Can any waveshape be clamped?

2. With regard to the value of C, what are the frequency considerations to be taken into account when using a clamper circuit?

The transistor switch

OBJECTIVES

After studying this unit, the student will be able to

- Describe the operation of a transistor switch
- Design and analyze a transistor switch circuit
- Discuss why speed-up capacitors are required in certain switch applications

SWITCHING CHARACTERISTICS AND THE LOAD LINE

The transistor is often used in pulse circuitry as a switching device. This use of a transistor is one of its most important applications. The transistor is operated in one of two conditions or states when it is used as a switching device or switch. Figure 4-1 illustrates these states which are known as the conducting and nonconducting states.

Figure 4-1a shows a transistor in the reverse bias condition. The base of the transistor

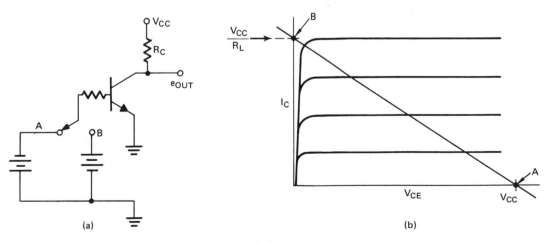

FIG. 4-1 THE TRANSISTOR SWITCH

is connected to the negative terminal of the bias battery (point A), and the transistor emitter is at ground. This means that the NPN transistor will not conduct. Therefore, any voltage developed at its collector (e_{out}) will be equal to V_{CC} and the transistor is acting as an open circuit. This point can be plotted on a load line as shown in figure 4-1b. Since the transistor is not conducting, there is zero collector current flowing. For an ideal situation, then, V_{CC} appears across the collector-emitter junction (V_{CE}).

If the base connection is changed from point A to point B, the transistor is now forward biased. As a result, the transistor is turned on and current will flow. The current flows through R_C and creates a voltage drop across R_C which is equal to V_{CC}. Therefore, the transistor has zero voltage across its emitter-collector junction and the transistor appears as a short circuit. The voltage e_{out} is equal to zero.

The reverse or forward biasing of a simple transistor circuit, such as the one shown in figure 4-1, provides the capability of switching large amounts of power.

PROBLEM 1

An ideal transistor switch will be analyzed in this problem. The term ideal is used because:

1. Actual transistors develop some voltage V_{BE}
2. Actual transistors develop some voltage V_{CE} even when fully saturated.
3. A square wave has some rise and fall time associated with it.

Therefore, this problem requires the designing of a transistor switch that will have the following circuit parameters:

(a) e_{out} must have a peak value of 30 volts
(b) e_{in} must be a rectangular wave varying from zero to five volts (+0 V to + 5 V)
(c) $I_{C\ max}$ = 30 mA (no load connected across the output terminals)
(d) The transistor used will be an NPN transistor with an h_{FE} of 60
(e) I_{CBO}, $V_{CE\ sat}$, and V_{BE} are all assumed to be zero

The circuit for this transistor switch is shown in figure 4-2. Since e_{out} must be 30 volts

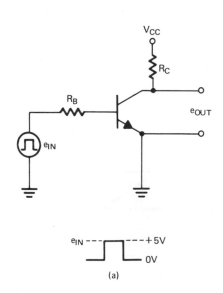

(a)

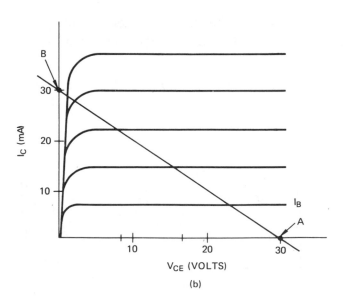

(b)

FIG. 4–2

peak (given in parameters for the circuit), V_{CC} must be 30 volts. The student should note that in a transistor switch circuit, the value of V_{CC} determines the amplitude of the output voltage.

Although the collector curves and the load line are not required to solve this problem, they are included so that the student will become familiar with the use of these valuable tools for circuit analysis. Refer to figure 4-2b where one point can be plotted for the load line. When the switch is open, the transistor is not conducting and V_{CE} will be equal to V_{CC} and I_C will be zero. The point at which these conditions are met is point A on the load line. The other point of the load line can be plotted directly from the circuit parameters. Point B occurs at $I_{C\,max} = 30$ mA.

If the circuit is to function properly, the values of R_C and R_B must be determined correctly.

R_C is found by dividing the maximum voltage (V_{CC}) by the maximum current (I_C). Recall that when the switch is closed and the transistor is conducting, the maximum voltage will be dropped across R_C. As a result, the maximum current will be flowing through R_C.

$$R_C = \frac{V_{CC}}{I_C}$$
$$R_C = \frac{30\text{ V}}{30\text{ mA}}$$
$$R_C = 1\text{ k}\,\Omega$$

The value of R_B is found by dividing the voltage drop across R_B by the current flowing through R_B. Since the voltage drop across R_B is e_{in} and the current through R_B is I_B,

$$R_B = \frac{e_{in}}{I_B}$$

The value of e_{in} is known but I_B is unknown. However, the following is true of I_B:

$$I_B = \frac{I_C}{h_{FE}}$$

Therefore,
$$I_B = \frac{30\text{ mA}}{60}$$
$$I_B = 0.5\text{ mA}$$
and
$$R_B = \frac{5\text{ V}}{0.5\text{ mA}}$$
$$R_B = 10\text{ k}\,\Omega$$

Now that the resistor values for this transistor circuit are known, it can be assumed that when the input pulse is applied, the transistor will be forward biased. As a result, the transistor will conduct at saturation and appear as a closed switch. On the other hand, if the input pulse is removed, there will be no base current and the transistor will not conduct. Therefore, there will be no collector current, no voltage drop across R_C, and the transistor will act as an open switch. The full applied voltage will appear across the collector-emitter junction of the transistor.

Explain the term saturation. (R4-1)

The ideal transistor switch will dissipate power only during the time it requires to go from one state to the other. While the transistor is in the conducting state (closed switch), the voltage drop across the transistor is zero. Although the current is at a maximum value at this time, the power is zero as shown in the following expression:

P = E x I
P = 0 x 30 x 10⁻³
P = 0

Similarly, when the transistor is not conducting and the switch is open, the voltage is at a maximum but the current is zero. As a result, the power is still zero.

THE PRACTICAL TRANSISTOR SWITCH

The concept of the ideal transistor was used to introduce the discussion of the transistor switch. In practical circuits, however, transistors are not ideal. A transistor does experience a voltage drop across it even when it is conducting at saturation. An important feature of actual transistor switches is the fact that the nonideal transistor cannot turn on in zero time. It was assumed in Problem 1 that once the input pulse is applied, current starts to flow immediately and the transistor turns on (e_{out} goes to zero). Another assumption made in Problem 1 is that when the transistor is on (conducting) and the input pulse is removed, the output voltage, e_{out}, immediately returns to the value of V_{CC}. However, this is not the case for practical switch circuits.

Figure 4-3 shows the practical switching characteristics of a transistor switch. To understand what happens when a pulse is applied to the input of the transistor switch, the student should refer to figures 4-3 and 4-4.

Figure 4-4a shows an enlarged view of the emitter-base junction of the transistor. (Recall that the p material is the base and the n material is the emitter.) At the time the input pulse is negative, this junction is reverse biased. The n and p materials are relatively low resistance areas and the region at the junction (space charge region) can be thought of as an insulator between the n and p regions. This situation amounts to a capacitor at the emitter-base junction. The value of this capacitor depends on the width of the space charge region; the width of the space charge region depends on the amount of reverse voltage present.

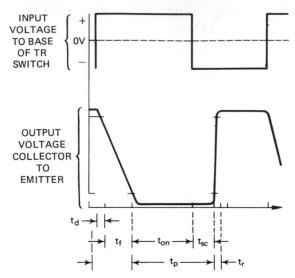

FIG. 4-3 WAVEFORMS OF A TRANSISTOR SWITCH

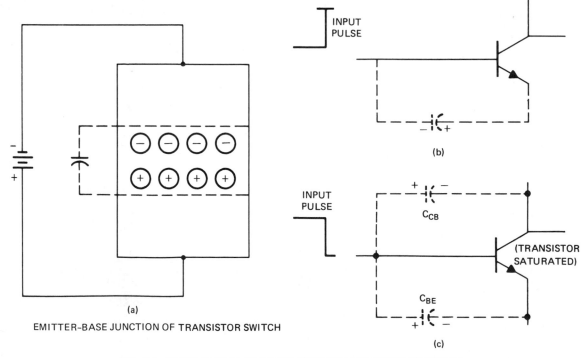

EMITTER-BASE JUNCTION OF TRANSISTOR SWITCH

FIG. 4-4 EMITTER-BASE JUNCTION CAPACITANCES IN THE TR SWITCH

Figure 4-4b is a representation of the capacitance at the junction. The figure shows the capacitor charge prior to the change to the positive input pulse. When the pulse goes positive, the capacitor must discharge to zero before it can begin to charge in the opposite direction. During this time, the transistor cannot turn on because the base-emitter junction is still reverse biased.

Once the transistor turns on, the emitter-base junction capacitance is charged as shown in figure 4-4c. Since the base-collector junction also has some capacitance, its charge is shown in figure 4-4c as well. When the input pulse is removed, both of these capacitors will attempt to return to zero. The capacitor action will cause the transistor to remain in the on condition for some time after the input pulse is removed. Note in figure 4-3 that the storage time of the junction capacitors, t_{sc}, is the time required by the capacitors to discharge enough to turn the transistor off.

These delay times are critical in very fast pulse circuitry where a number of circuits must respond in sequence and on time. One method of overcoming the delay in the turn-on time is to increase the base current. If the base current is doubled, the fall time (t_f of figure 4-3) is reduced by about one-third; if the base current is tripled the fall time is reduced by about one-fifth. Of course, any increase in the base current causes an increase in the charge built up on the junction capacitors. As a result, the turn-off time is delayed by an even greater amount. The balance of this unit covers a method by which faster switching can be accomplished.

THE USE OF SPEED-UP CAPACITORS IN TRANSISTOR SWITCHING

To obtain faster switching times, two conditions must be satisfied:

1. An increase in the initial current flow when an input pulse is applied, and

2. A decrease or elimination of the storage charge on the junction capacitors so that the transistor can turn off more quickly.

One method of achieving both of these objectives is shown in figure 4-5. The capacitor connected across R_B is called a speed-up capacitor. When the input pulse first goes positive, the components of the base current are the current needed to charge the capacitor and the current through R_B.

As a result, the current through the base is greatly increased.

The capacitor appears initially as a short and current through it is at a maximum value which is limited only by the base-emitter resistance of the transistor. As the capacitor gradually charges and the current decreases to the value of the current through R_B, the transistor switch turns on faster.

When the input pulse goes negative, or back to zero, the base sees a high negative charge from the speed-up capacitor. The base is almost immediately reverse biased, thus cutting off the transistor.

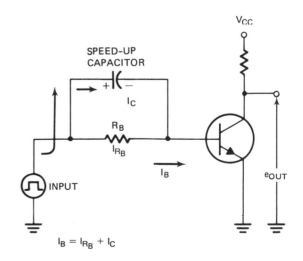

FIG. 4-5 THE SPEED-UP CAPACITOR IN THE TR SWITCH

EXTENDED STUDY TOPICS

1. This unit gives a rule of thumb concerning the effect of doubling or tripling the base current with regard to reducing the delay time. Verify this assumption mathematically.

   Hint: $1 - \epsilon^{-tf/\tau} = 80\%$ of total fall time

The unijunction oscillator

OBJECTIVES

After studying this unit, the student will be able to:

- Describe the operation of a unijunction transistor.
- Design and construct a UJT relaxation oscillator

THE UJT

The *unijunction transistor* (UJT) is a three-lead semiconductor device. UJTs are used in numerous types of circuits such as multivibrators, sawtooth generators, and time-delay circuits.

Figure 5-1 shows the structure of a unijunction transistor. The device has three leads but only one *p-n* junction. Thus, the device is described as a single (uni) junction device. The UJT consists of a bar of *n*-type silicon to which two leads, A and B, are bonded on the longitudinal axis. These connections are resistive (or ohmic) connections and are nonrectifying. An ohmmeter reading taken between A to B will give the same reading as one taken between B to A. A piece of *p*-type material, figure 5-1b, is placed just past the center of the *n*-type material toward the A lead. The junction of this *p*-type material and the main bar of *n*-type material forms the *p-n* junction of the device. Figure

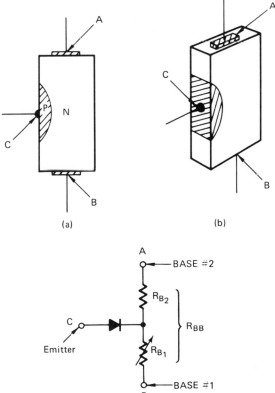

FIG. 5-1 CONSTRUCTION OF THE UNIJUNCTION TRANSISTOR (UJT)

5-1c shows the equivalent circuit for the UJT. Note that the total resistance of the device, the resistance from A to B, is shown as two resistors. This situation is not true in actual practice, but is used here to simplify the presentation of the theory of operation for the device. Since the *p-n* junction (equivalent diode) is placed closer to point A than to point B, the resistance R_{B_2} is slightly less than the resistance of R_{B_1}. The total resistance of the UJT (and the only resistance that can be measured) is R_{BB} and,

$$R_{BB} = R_{B_1} + R_{B_2} \qquad \text{Eq. 5.1}$$

Note that R_{B_1} is shown as a variable resistor in figure 5-1. The reason for designating R_{B_1} in this manner will be covered later in this unit. The UJT operates when the diode is forward biased, figure 5-2. If a bias battery is placed as shown, the voltage from this battery will divide proportionally across R_{B_1} and R_{B_2}. If an external voltage is now placed between points C and B, the diode will be forward biased when this voltage exceeds the voltage across R_{B_1}. Since R_{B_1} cannot be measured (or the voltage drop across it), it will require a trial and error procedure to find the exact voltage required to forward bias the diode. Each device will vary slightly in its requirements; thus, a trial and error approach to finding the voltage will be very time consuming. UJT manufacturers have simplified the task by specifying a value equal to the ratio

$$\frac{R_{B_1}}{R_{BB}}$$

This ratio is known as the intrinsic standoff ratio and is expressed using the Greek letter η (eta).

$$\eta = \frac{R_{B_1}}{R_{BB}} \qquad \text{Eq. 5.2}$$

Equation 5.2 can be rearranged to the form

$$R_{B_1} = \eta \times R_{BB} \qquad \text{Eq. 5.3}$$

Since $R_{BB} = R_{B_1} + R_{B_2}$

$$\eta = \frac{R_{B_1}}{R_{B_1} + R_{B_2}} \qquad \text{Eq. 5.4}$$

In terms of voltage drops, the ratio can be expressed as:

$$\eta = \frac{V_{B_1}}{V_{BB}} \qquad \text{Eq. 5.5}$$

PROBLEM 1

For a UJT with an η specified by the manufacturer as 0.62, and whose measured R_{BB} is 7.7 kΩ find V_{B_1} and V_{B_2}. The applied voltage, V_{BB}, is 12 volts.

From equation 5.5,

$$\eta = \frac{V_{B_1}}{V_{BB}}$$

$$\text{or} \quad V_{B_1} = V_{BB} \cdot \eta$$

Therefore, $V_{B_1} = 0.62 \times 12$ V

$$V_{B_1} = \textbf{7.44 volts}$$

Using Kirchhoff's voltage laws, the following expression can be written,

$$V_{BB} = V_{B_1} + V_{B_2}$$

$$\text{and} \quad V_{B_2} = V_{BB} - V_{B_1}$$

Therefore, $V_{B_2} = 12$ V – 7.44 V

$$V_{B_2} = \textbf{4.36 volts}$$

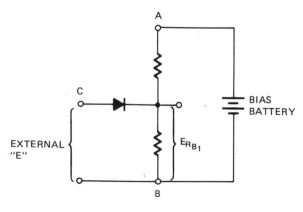

FIG. 5-2   BIAS CONFIGURATION OF A UJT

A certain UJT has an intrinsic standoff ratio, η, of 0.64 and a measured interbase resistance (resistance from base 1 to base 2) of 8 k ohms. What is the voltage on R_{B1} and R_{B2} if V_{BB} is 12 volts? (R5-1.)

In a typical unijunction transistor, is the n-type material silicon or germanium? (R5-2)

NEGATIVE RESISTANCE

Only the static resistive characteristics of the UJT have been covered in this unit to this point. What happens when the diode is forward biased? Based on the student's knowledge of semiconductors and *p-n* junction theory, it should be apparent that forward biasing the *p-n* junction causes holes to be injected into the *n*-type material. Because of these holes, a greater current will flow which, in turn, causes more holes to be injected in the *n*-type material. As a result, the resistance of R_{B1} is lowered, thus accounting for showing it as a variable resistor in figure 5-1. The decrease in the resistance of R_{B1} causes the voltage across R_{B1} to decrease according to Ohm's Law. The decreased voltage drop further forward biases the diode with a resulting additional increase in the current. Therefore, as the emitter current increases, the emitter voltage will decrease. Since the diode resistance drops to a very low value and

R_{B1} is decreasing, the voltage also decreases, figure 5-3. Note in figure 5-3 that as the emitter voltage increases from zero to some

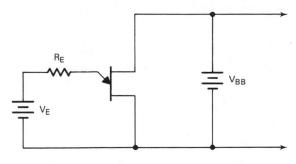

FIG. 5-4   CIRCUIT TO ESTABLISH V_E FOR UJT

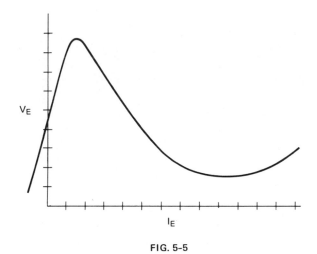

FIG. 5-5

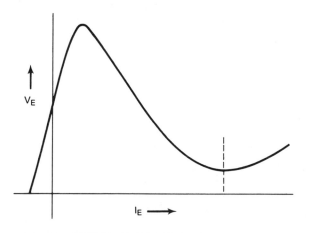

FIG. 5-3   EMITTER VOLTAGE/CURRENT CHARACTER-
ISTICS OF THE UNIJUNCTION TRANSISTOR

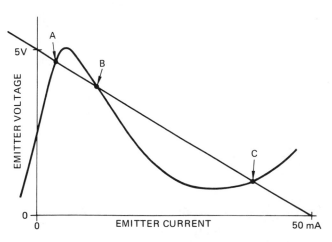

FIG. 5-6   LOAD LINE FOR BISTABLE UJT CIRCUIT

37

positive value, initially there is a small negative (reverse) emitter current. The positive (forward) emitter current starts to flow when the emitter to base one voltage reaches a value great enough to forward bias the *p-n* junction. Once this emitter current begins to flow, the base resistance R_{B_1} decreases with the result that V_E begins to decrease. Voltage V_E continues to decrease, even though I_E is increasing, until R_{B_1} reaches its lowest value which is called the *saturation resistance,* R_{SAT}. This portion of the curve in figure 5-3, page 37, is called the *negative resistance region* because of the fact that as the current increases, the voltage decreases.

UJT CIRCUIT ANALYSIS

The experimental circuit in figure 5-4, page 37, can be used to analyze the various operating modes of a simple unijunction transistor circuit. Figure 5-5, page 37, shows the characteristic curve for the UJT used in the circuit of figure 5-4, page 37. (Recall that a set of UJT emitter current versus emitter voltage curves is drawn for constant value of V_{BB}.) For the circuit of figure 5-4, V_{BB} is assumed to be 10 volts. If values are assigned to R_E and V_E, R_E = 100 ohms and V_E = 5 volts, a load line can be drawn on the characteristic curve as shown in figure 5-6, page 37.

The load line crosses the characteristic curve at three different points. Two of the points are in positive resistance regions (A and C), and the third point (B) is in the negative resistance region of the curve. For a load line as shown in figure 5-6, the circuit generally will not operate in the region around point B. Point B is in the unstable or negative resistance region. Thus, circuit operation will shift easily to either point A or point C since these points are in the positive resistance region of the characteristic curve. Once UJT operation is established at either point A

or point C, it will remain at that point until it is switched to another point on the curve. This stability is in contrast to operation at point B where any slight change in the circuit parameters will cause the circuit operation to change. For the load line shown in figure 5-6, the circuit will have two stable states. As a result, this type of circuit is called a *bistable circuit*. A UJT circuit having a load line as shown in figure 5-6· is known as a *bistable (flip-flop) multivibrator*.

THE RELAXATION OSCILLATOR

The UJT can be used in the construction of another type of multivibrator known as the *relaxation oscillator* or as the *astable* or *free-running multivibrator*.

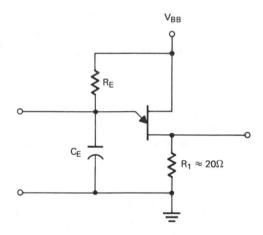

FIG. 5-7   CIRCUIT FOR UJT OSCILLATOR

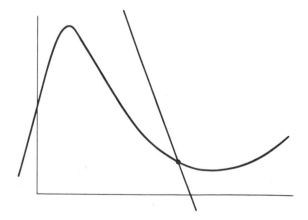

FIG. 5-8   LOAD LINE FOR UJT RELAXATION OSCILLATOR CIRCUIT

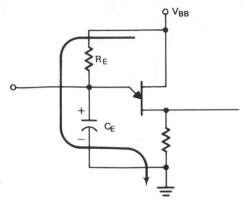

TIME OF CHARGE = R_E C_E

(a) CHARGE PATH OF C_E

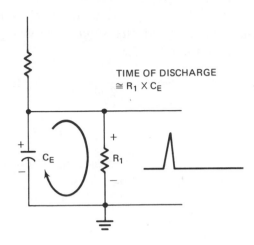

TIME OF DISCHARGE
$\cong R_1 \times C_E$

(b) DISCHARGE PATH OF C_E

FIG. 5-9

Two conditions must be met before a relaxation oscillator circuit can be constructed. The first condition is that for the UJT oscillator circuit shown in figure 5-7, page 38, the value of R_E must be selected to insure that the load line crosses the characteristic curve at one point and only one point, figure 5-8, page 38. In addition, that point must be in the negative resistance region of the curve. To insure that the circuit will be unstable, the second condition is that a capacitor (C_E) must be connected from the emitter to base 1. The resistor R_1 is added to develop a signal voltage at base 1.

The following steps present the sequence of operation of this oscillator circuit.

1. When power is applied, the capacitor begins to charge to the supply voltage (V_{BB}) through R_E. The time rate of this charge is equal to the RC time constant of R_E and C_E.

2. The capacitor charges to a voltage value which will forward bias the *p-n* junction of the UJT.

3. When this voltage value is reached, the *p-n* junction of the UJT will short circuit. This short circuit plus the small resistance of R_1 are now across the capacitor.

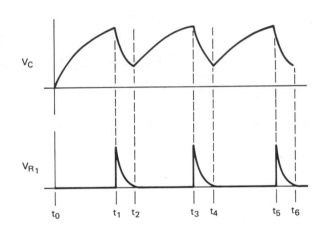

CHARGE TIME EQUALS $t_0 - t_1, t_2 - t_3, t_4 - t_5$

DISCHARGE TIME EQUALS $t_1 - t_2, t_3 - t_4, t_5 - t_6$

FIG. 5-10   WAVESHAPES OF UJT
RELAXATION OSCILLATOR

4. The capacitor has a low resistance path for discharge, figure 5-9. As the capacitor discharges, current is produced through R_1. This current produces a short spike of output voltage across R_1.

5. When the capacitor is discharged, it will begin to recharge through R_E and the cycle will be repeated.

Figure 5-10 shows the relationship of the charge voltage across C_E and the voltage produced across R_1.

The frequency of an oscillator circuit (the frequency of the spikes across R_1) is found by using Eq. 5.6.

$$f = \frac{1}{(R_E)\,(C)\,\ln\left(\frac{1}{1-\eta}\right)} \qquad \text{Eq. 5.6}$$

PROBLEM 2

A UJT oscillator circuit has the following values: $\eta = 0.64$, $C = 0.01\ \mu F$, and $R_E = 2\ k$ ohms. Calculate the frequency of the oscillations.

$$f = \left[\frac{1}{(2 \times 10^3)\,(0.01 \times 10^{-6})\,\ln\left(\frac{1}{0.36}\right)}\right]$$

$$f = \left[\frac{1}{(2 \times 10^3)(0.01 \times 10^{-6})\,\ln 2.77}\right]$$

$$f = \left[\frac{1}{(2 \times 10^3)(0.01 \times 10^{-6})(1.02)}\right]$$

$$f = 48.9\ k\ Hz$$

An estimate of this frequency can be obtained by the use of Eq. 5.7

$$f = \frac{1}{R_E \times C_E} \qquad \text{Eq. 5.7}$$

Substituting the values of Problem 2 in this equation yields:

$$f = \frac{1}{(2 \times 10^{-3})(0.01 \times 10^{-6})}$$

$$f = 50\ k\ Hz$$

Either equation may be used depending on the degree of accuracy required for the circuit.

The student should recall that resistor and capacitor values usually have a 10 or 20 percent tolerance range. Therefore, when performing calculations, the student should use measured values of resistance and capacitance. A UJT oscillator can be used as a time-delay circuit. In this situation, the time $t_n - t_{n-2}$ (figure 5-10), is of interest.

Since time $= \left(\frac{1}{f}\right)$,   Eq. 5.8

the equation for time delay can be expressed as

$$TD = \frac{1}{\dfrac{1}{R_E \times C_E \left[\ln\left(\frac{1}{1-\eta}\right)\right]}}$$

or   Eq. 5.9

$$TD = R_E \times C_E \left[\ln\left(\frac{1}{1-\eta}\right)\right]$$

Eq. 5.9 can be approximated to:

$$TD = R_E \times C_E \qquad \text{Eq. 5.10}$$

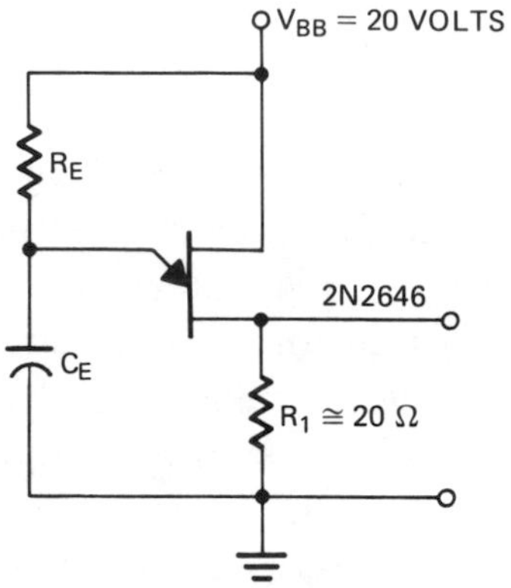

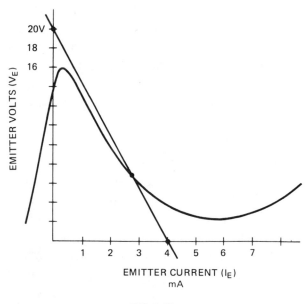

FIG. 5-11

FIG. 5-12

PROBLEM 3

Using the circuit of figure 5-11 and the load line of figure 5-12, determine

a. the value of R_E needed to produce the load line,
b. the value of η for the UJT shown, and
c. the capacitance value needed to produce a time delay of 2 seconds.

Draw a load line through V_{BB} to the left of the valley voltage point and crossing the curve at only one point.

Since $R_E = \dfrac{V_{BB}}{I_E}$,

$R_E = \dfrac{20\,V}{4\,mA} = 5\,k\,ohms$

Since $\eta = \dfrac{V_{B1}}{V_{BB}}$, and $V_{B1} = V_P$

$\eta = \dfrac{V_P}{V_{BB}} = \dfrac{16\,V}{20\,V} = 0.8$

$TD = R_E \times C_E \left[\ln \left(\dfrac{1}{1-\eta} \right) \right]$

$2\,sec = 5\,k\,\Omega \times C_E \left[\ln \left(\dfrac{1}{1-0.8} \right) \right]$

$2\,sec = 5\,k\,\Omega \times C_E \ln 5$
$2\,sec = 5\,k\,\Omega \times C_E \times 1.61$
$2\,sec = (8.05 \times 10^3)(C_E)$
$\dfrac{2\,sec}{8.05 \times 10^3} - C_E$

$C_E = 249\,\mu F$

LABORATORY EXERCISE 5-1:   RELAXATION OSCILLATOR

PURPOSE

- To provide the student with experience in designing a relaxation oscillator.

PROCEDURE

A. Using a 2N2646 UJT, design a relaxation oscillator circuit similar to that of figure 5-11. Refer to the specification sheet for this device and examine its electrical characteristics.

B. Sketch the waveshapes for the circuit as shown in figure 5-10.

C. Determine the reasons for any differences between the measured and calculated values of the frequency, time delay, and the values of R_E and C_E.

EXTENDED STUDY TOPICS

1. Construct a test circuit to measure the following UJT characteristics:

   a. V_P
   b. I_P
   c. V_V
   d. I_V

   e. η
   f. R_{B_1}
   g. R_{B_2}
   h. R_{BB}

2. Plot the graph of the characteristic curves of the UJT tested in Topic 1. (Use the 2N2646 UJT from the Laboratory Exercise.)

6

The Schmitt trigger

OBJECTIVES

After studying this unit, the student will be able to

- Design a Schmitt trigger circuit
- Use the Schmitt trigger to produce a rectangular wave of a desired duration.

The Schmitt trigger is another of several circuits which are used to produce nonsinusoidal waves. The Schmitt trigger is often used as a voltage level (amplitude) detector. Figure 6-1a is a block diagram illustrating the basic operation of the Schmitt trigger. When the Schmitt trigger receives a sine-wave input, a rectangular wave is produced at the output. The output pulse has a duration which is less than one-half of the cycle time T for the input sine wave. In other words, the pulse duration of the output rectangular wave *is always* less than one-half the time of one cycle of an input sine wave. The conditions leading to this value for the output pulse duration will be determined by analyzing the Schmitt trigger circuit of figure 6-2, page 43.

The Schmitt trigger consists of two transistor switches (refer to Unit 4 for an analysis of transistor switch operation). Initially, the

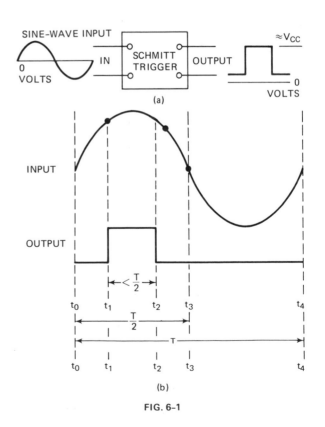

FIG. 6-1

two transistors are biased so that Q_1 is turned off and Q_2 is turned on. When Q_2 is turned on, the voltage e_{out} is approximately equal to the voltage drop across R_E. When the input voltage level of the sine wave reaches a level high enough to turn Q_1 on, the voltage applied to the base of Q_2 decreases since the collector voltage of Q_1 is the base voltage for Q_2. As the base voltage Q_2 decreases, Q_2 becomes reverse biased (the voltage is less than emitter voltage) and Q_2 turns off. When Q_2 turns off, its collector voltage increases toward the value of V_{CC}. As the input sine wave begins to decrease, it reaches the value at which Q_1 turns off. When Q_1 turns off, its collector voltage increases, and the base voltage of Q_2 increases until Q_2 is turned on again with the result that the output voltage again decreases to V_{RE}.

During the operation of the Schmitt trigger, note that the output voltage pulse, ($t_1 - t_2$), can occur only during a portion of the positive half of the input sine wave. The point on the input sine wave at which Q_1 turns on is called the *upper trigger point* (UTP); similarly, the point on the input sine wave at which Q_1 turns off is called the *lower trigger point* (LTP).

For a 2 k Hz sine-wave input, what is the maximum pulse width of the output rectangular wave? (R6-1)

For what application is a Schmitt trigger particularly suited? (R6-2)

CALCULATING THE UPPER TRIGGERING POINT

The Schmitt trigger circuit of figure 6-2 can be used to calculate the voltage level required to turn on Q_1. This voltage will be the higher of the two voltage levels of the input wave which cause the changes in the operation of the Schmitt trigger. When Q_1 is in the off state, Q_2 is in the on state. If Q_2 is on and conducting, there will be a voltage drop across

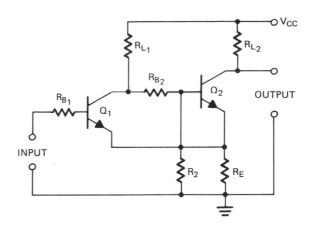

FIG. 6-2   THE SCHMITT TRIGGER

R_E due to the emitter current. This current is equal to the collector current plus the base current of Q_2. If Q_1 is to be turned on, its base must be more positive than its emitter. In other words, the input voltage must reach a level greater than V_{RE} to turn on Q_1. The upper trigger level then is equal to V_{RE}.

$$UTP = V_{RE} \qquad \text{Eq. 6.1}$$

$$\text{Since } V_{RE} = i_E R_E, \qquad \text{Eq. 6.2}$$

where i_E is the value for Q_2,

the upper trigger point can be determined for the Schmitt trigger once i_E is known.

For any given Schmitt trigger, the values of the upper trigger point and the lower trigger point are determined by the circuit use. Therefore, it is necessary only to calculate the value of R_E. Since $I_c \cong I_E$, the selection of values for V_{CC}, I_C and UTP means that the circuit can be made to trigger at that point once the proper value of R_E is known.

Why is the assumption $I_C \cong I_E$ valid? (R6-3)

For the UTP,

$$(R_{L2} + R_E) = \frac{V_{CC}}{I_C} \qquad \text{Eq. 6.3}$$
$$\text{and } V_{RE} = UTP$$

Therefore,

$$V_{RE} = V_{CC}\left(\frac{R_E}{R_{L_2} + R_E}\right) \qquad \text{Eq. 6.4}$$

so $R_E = (V_{RE}/V_{CC})(R_{L_2} + R_E)$

$$R_E = \left(\frac{V_{RE}}{V_{CC}}\right)\left(\frac{V_{CC}}{I_C}\right)$$

$$R_E = \frac{V_{RE}}{I_C}$$

and $R_E = \dfrac{UTP}{I_C} \qquad \text{Eq. 6.5}$

For a circuit where the required upper triggering point is 5 volts and the desired collector current at Q_2 is 20 mA, calculate R_E. (R6-4)

CALCULATING THE LOWER TRIGGERING POINT

Once Q_1 is turned on and Q_2 is turned off, the cycle of the Schmitt trigger is completed by turning off Q_1 once more. When Q_1 is on and Q_2 is off, the volage across R_E changes and is given by the expression:

$$V_{R_E} = (I_{E_{Q1}})(R_E)$$

where $I_{E_{Q1}} = I_{C_{Q1}}$

If Q_1 is to be turned off, the input voltage (LTP) must be equal to $V_{R_E} = (I_{E_{Q1}})(R_E)$. The lower triggering point is selected by specifying the input voltage at which Q_1 turns off. Once this value is selected, the proper value of V_{R_E} is found by determining the value of R_{L_1}:

$$V_{RE} = \left(\frac{R_E V_{CC}}{R_E + R_{L_1}}\right) \qquad \text{Eq. 6.6}$$

and $R_{L_1} = \left(\dfrac{R_E V_{CC}}{LTP}\right) - R_E$

CALCULATING THE VALUES OF THE BASE RESISTORS

To complete the design of the Schmitt trigger circuit, it is necessary to determine the values of the two base resistors. The value of R_{B_1} must be less than the product of the hfe of Q_1 times the value of R_E. To insure that an adequate value of R_{B_1} is selected, a safety factor is used. Eq. 6.7 shows a safety factor of 10, but the factor may be as low as 2 and still provide an adequate value of R_{B_1}.

$$R_{B_1} = \left(\frac{hfe \times R_e}{10}\right) \qquad \text{Eq. 6.7}$$

Since R_{B_2} is in the circuit when Q1 is off, figure 6-3, R_{B_2} can be calculated from Eq. 6.8.

$$R_{B_2} = \frac{E_{R_{B_2}}}{I_{R_{B_2}}} \qquad \text{Eq. 6.8}$$

where $I_{R_{B_2}} = I_{B_2} + I_{R_2} \qquad \text{Eq. 6.9}$

and $I_{B_2} = \dfrac{I_{C_2}}{hfe}$

$$I_{R_2} = \frac{I_{C_2}}{10}$$

Therefore, $I_{R_{B_2}} = \dfrac{I_{C_2}}{hfe} + \dfrac{I_{C_2}}{10}$

with $E_{R_{B_2}} = V_{CC} - E_{R_{L_1}} - E_{R_2} \qquad \text{Eq. 6.10}$

$$E_{R_{B2}} = V_{CC} - (I_C)(R_{L_1}) - \frac{I_{C_2}}{10}(R_2)$$

and $R_2 = \dfrac{E_{R_2}}{I_2} = \dfrac{LTP}{\left(\dfrac{I_{C_2}}{10}\right)}$

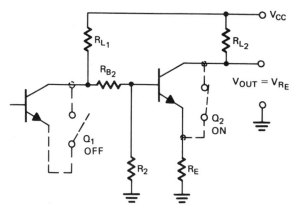

FIG. 6-3 CIRCUIT FOR SCHMITT TRIGGER CALCULATIONS

With the information given to this point in the unit, it is possible to design a Schmitt trigger circuit as demonstrated in the following problem.

PROBLEM 1

Using the circuit of figure 6-4, a Schmitt trigger is to be designed to meet the following circuit specifications:

- $V_{CC} = 12$ volts
- Maximum $I_C = 6$ mA
- UTP = 6 volts
- LTP = 3 volts

The transistors used in the circuit are silicon NPN transistors with hfe min = 20. Assume that the transistors are ideal and that $I_{CBO} = 0$ (that is, the collector to base leakage current is zero). Also assume that the junction voltages are equal to zero.

Calculate the value of each resistor required to provide the proper biasing so that the Schmitt trigger will operate at the UTP and LTP values.

1. With no input signal, the circuit must be biased so that Q2 is on and Q1 is off. Use figure 6-5 to calculate the combined value of R_E and R_{L_2}.

$$R_E + R_{L_2} = \frac{V_{CC}}{I_{C_2}}$$

$$= \frac{12\text{ V}}{6\text{ mA}}$$

$$R_E + R_{L_2} = 2\text{ k ohms}$$

2. If the junction voltage of the transistors is assumed to be zero, then

$$\text{UTP} = V_{R_{E_2}}$$

where $V_{R_{E_2}}$ is the voltage across R_E when Q2 is on. V_{R_E} can be determined using the voltage divider principle.

$$V_{R_{E_2}} = \frac{R_E \; V_{CC}}{(R_{L_2} + R_E)}$$

Therefore, $V_{R_{E_2}} = \left(\frac{R_E \; V_{CC}}{2\text{ k }\Omega}\right)$

and UTP $= \left(\frac{(R_E) \; (12\text{ V})}{2\text{ k }\Omega}\right)$

$$6\text{ V} = \left(\frac{(R_E) \; (12\text{ V})}{2\text{ k }\Omega}\right)$$

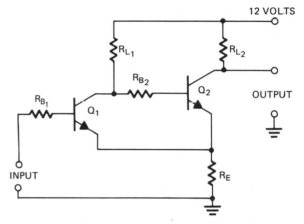

FIG. 6-4

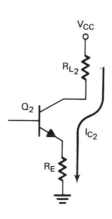

FIG. 6-5   COLLECTOR CURRENT WITH Q_2 ON

Solve for R_E:

$$R_E = \left[\frac{(6\text{ V})(2\text{ k}\Omega)}{12\text{ V}}\right]$$

$$R_E = 1\text{ k}\Omega$$

3. Since $(R_{L_2} + R_E) = 2\text{ k}\Omega$, R_{L_2} can be found.

$$R_{L_2} + 1\text{ k}\Omega = 2\text{ k}\Omega$$

$$R_{L_2} = 2\text{ k}\Omega - 1\text{ k}\Omega$$

$$R_{L_2} = 1\text{ k}\Omega$$

4. Once Q1 is turned on, an equivalent circuit can be drawn as shown in figure 6-6. To turn Q1 off and complete the cycle of operation of the Schmitt trigger, the input voltage must be lowered to a point equal to the potential across R_E at this time. This voltage level is the lower trigger point (LTP).

The voltage divider method is again applied to obtain the following equation.

$$V_{R_{E_1}} = \left(\frac{R_E\, V_{CC}}{R_E + R_{L_1}}\right)$$

The equation for $V_{R_{E_1}}$ can be solved for R_{L_1}

To solve for R_{L_1} we will transpose this equation to read

$$R_{L_1} = \left[\frac{R_E\, V_{CC}}{V_{R_E}}\right] - R_E$$

where V_{R_E} = LTP = 3 V

Therefore,

$$R_{L_1} = \frac{(1\text{ k}\Omega)(12\text{ V})}{3\text{V}} - 1\text{ k}\Omega$$

$$R_{L_1} = (1\text{ k}\Omega)(4) - 1\text{ k}\Omega$$

$$= 4\text{ k}\Omega - 1\text{ k}\Omega$$

$$R_{L_1} = 3\text{ k}\Omega$$

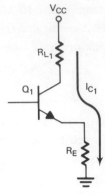

FIG. 6-6 COLLECTOR CURRENT WITH Q_1 ON

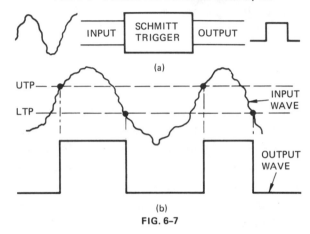

FIG. 6-7

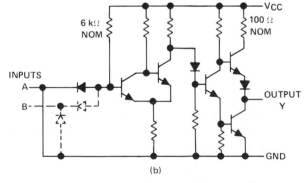

FIG. 6-8 MODEL SN7414 SCHMITT TRIGGER

5. To bias Q1 properly, the value of R_{B_1} must be less than R_E times hfe min. Therefore,

$$R_{B_1} = \left[\frac{(hfe\ min)\ (R_E)}{10} \right]$$

$$= \frac{(20)\ (1\ k\ \Omega)}{10} = 2\ k\ \Omega$$

(The expression can also be written using the less conservative safety factor:

$$R_{B_1} = \left[\frac{(hfe\ min)\ (R_E)}{2} \right]$$

6. R_{B_2} can be found by the use of the following expression.

$$R_{B_2} = \frac{E_{R_{B_2}}}{I_{R_{B_2}}}$$

Substitute Eq. 6.9 and Eq. 6.10 into the above equation to obtain:

$$R_{B_2} = \frac{V_{CC} - E_{R_{L_1}} - E_{R_{L_2}}}{I_{B_2} + I_{R_2}}$$

$$= \left[\frac{12\ V - (9\ V)}{2 \left(\frac{I_{C_2}}{hfe} \right)} \right]$$

$$= \frac{3\ volts}{(2)\ (0.3 \times 10^{-3})} =$$

$$R_{B_2} = 5\ k\ \Omega$$

USES OF THE SCHMITT TRIGGER

The Schmitt trigger circuit will produce a rectangular wave of constant amplitude and varying pulse width when a sine wave of varying amplitude is present at its input. The Schmitt trigger may be used as a pulse shaper as shown in figure 6-7, page 46. The figure shows that a noisy input pulse (or signal) is converted to a clean output pulse with a constant amplitude. Since the Schmitt trigger circuit uses only the upper and lower

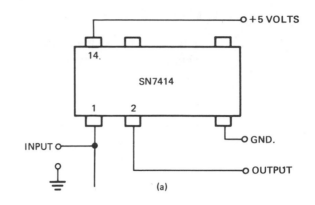

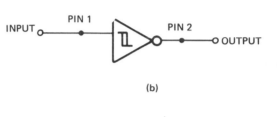

FIG. 6-9

triggering points (voltages), the shape of the input signal is unimportant.

INTEGRATED CIRCUIT SCHMITT TRIGGERS

Schmitt trigger circuits are available in integrated circuit (IC) form, such as the Texas Instrument Model SN7414 Schmitt trigger. This device is shown in block diagram form in figure 6-8, page 46. The balance of figure 6-8 shows the schematic diagram of each of the six circuits shown in the block diagram. The input for this device is pins 1-3-5-9-11-13. The output is at pins 2-4-6-8-10-12. V_{CC} is at pin 14 and ground is pin 7. V_{CC} is common to all six circuits as is the ground connection. These connections are omitted on the block diagram in figure 6-8 for clarity. Figure 6-9 shows a typical circuit using an IC Schmitt trigger. The connections to the integrated circuit are shown in figure 6-9a and the schematic diagram is shown in figure 6-9b. This device has a UTP of approximately 1.66 volts and an LTP of approximately 0.85 volts.

EXTENDED STUDY TOPICS

1. This unit investigated both discrete component and integrated circuit Schmitt trigger circuits. Construct the circuit of figure 6-4 and use this circuit to verify the theory outlined in this unit.

2. Describe several additional uses not covered in this unit for the Schmitt trigger circuit.

3. Construct the circuit of figure 6-9. Use various input waveforms (other than sine waves) to verify the theory of operation of this device.

Unit 7

Multivibrators

OBJECTIVES

After studying this unit, the student will be able to

- Describe and use the various types of multivibrator circuits
- Construct multivibrator circuits to produce the desired output waveforms.

Multivibrators are circuits which are capable of producing rectangular output waveshapes. Multivibrator circuits may be as simple as a mechanical switch connected to a load and dc supply, or they may be electronic switches which rely on the charging and discharging of RC time constant circuits to produce the desired output waveshapes. These electronic switches are usually regenerative transistor switch circuits. The Schmitt trigger circuit covered in Unit 6 is a form of multivibrator circuit. Three basic types of multivibrators will be covered in this unit:

- the *astable multivibrator* which is also known as a *free-running multivibrator*
- the *bistable multivibrator* which is also known as a *flip-flop* circuit (the bistable multivibrator will be examined in greater detail in the text DIGITAL LOGIC CIRCUITS)

- the *monostable multivibrator* which is also known as the *one-shot* or *time-delay multivibrator*

THE ASTABLE MULTIVIBRATOR

The astable multivibrator is shown in figure 7-1a, page 50. This type of multivibrator consists of two cross-coupled inverter circuits. The output of one inverter circuit is the input of the other circuit. The main feature of this circuit is that both of the inverters cannot operate in the saturated condition simultaneously. Therefore, when one inverter is conducting the other inverter circuit is cut off. Since Q1 and Q2 are constantly changing states (between off and on), this type of multivibrator does not have a stable state or condition but has two semi-stable states.

At time t_0 in the operation of the circuit in figure 7-1a, Q1 is conducting and Q2 is cut off. At this time, the collector of Q1 is at the ground potential and C_1 begins to charge toward V_{CC}. As C_1 charges, its voltage is coupled to the base of Q2. When this voltage is sufficient to forward bias the base-emitter junction of Q2 and thus turn on Q2, the collector voltage at Q2 will begin to drop. As Q2 continues to conduct, its collector voltage decreases rapidly toward the ground potential. The voltage drop at the Q2 collector is coupled to the base of Q1 through C_2. As a result, Q1 is reverse biased and turns off. The states of Q1 and Q2 have just been changed and now the process is reversed. In other words, C_2 begins to charge toward V_{CC}. Q1 once again is forward biased and turns on. This change in the states of Q1 and Q2 will continue with the result that the output at either the Q1 collector or the Q2 collector is a train of pulses as shown in figure 7-2, page 51.

The length of time that either Q1 or Q2 is off is determined by the RC time constants of R_1C_1 and R_2C_2. Recall from the study of time constants that the capacitor will charge to 50% of its maximum value of 0.69 TC. Therefore, the time during which each transistor is off is given by the following expressions:

Time Q1 off = 0.69R_2C_2 **Eq. 7.1**
Time Q2 off = 0.69R_1C_1 **Eq. 7.2**

In a symmetrical astable multivibrator, the total time for one period is twice the value of the time constant or,

Time = 1.38RC **Eq. 7.3**

The frequency of the output waveform is given by Eq. 7.4

$$F = \frac{1}{1.38RC}$$ **Eq. 7.4**

where

$R = R_1 = R_2$ and
$C = C_1 = C_2$

To produce varying output waveshapes, it is necessary to select the proper values of R_1, R_2, C_1, and C_2.

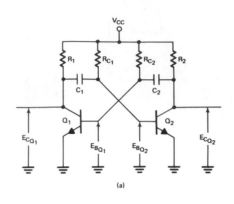

(a)

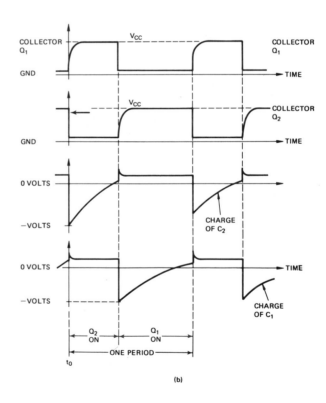

(b)

FIG. 7-1 THE ASTABLE MULTIVIBRATOR

To insure that the circuit functions properly, the transistor selected must have the proper value of hfe. The minimum value of hfe that will produce an output for the multivibrator is given by Eq. 7.5.

$$hfe_{min} = \frac{R_{Timing}}{R_{Load}} \qquad Eq.\ 7.5$$

where

$$R_{Load} = R_1 = R_2 \quad and$$
$$R_{Timing} = R_{C_1} = R_{C_2}$$

Why must a minimum hfe be known before an astable multivibrator can be constructed? (R7-1)

THE MONOSTABLE MULTIVIBRATOR

The monostable multivibrator has one stable state. Unlike the astable multivibrator which has two semistable states but no stable states, the monostable multivibrator has one stable state and one semistable state. Figure 7-3 shows that the monostable multivibrator is similar in construction to the astable multivibrator. In figure 7-1 for the astable multivibrator, both outputs are RC coupled back to the opposite inputs. For the monostable circuit, one output is RC coupled to the opposite input and the other output is resistively coupled to the opposite input.

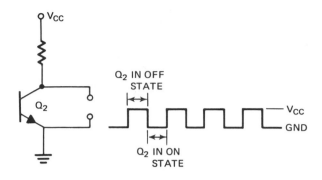

FIG. 7-2 OUTPUT WAVESHAPE AT COLLECTOR OF Q$_2$

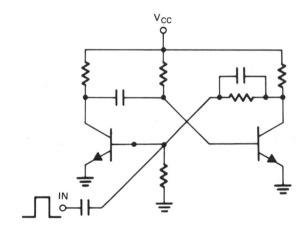

FIG. 7-3   THE MONOSTABLE (ONE-SHOT) MULTIVIBRATOR

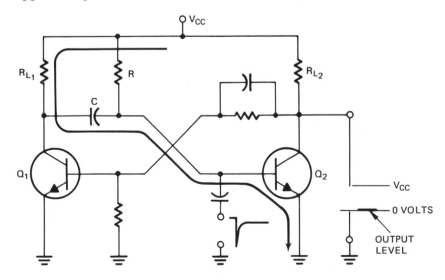

FIG. 7-4   THE MONOSTABLE MULTIVIBRATOR

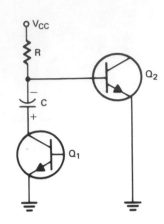

FIG. 7-5   CIRCUIT WHEN Q₁ TURNS ON

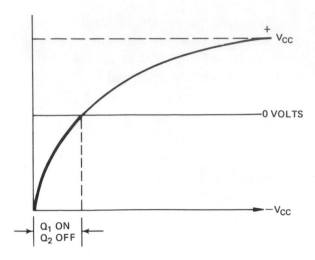

FIG. 7-6   CHARGE OF C FROM $-V_{CC}$ TO $+V_{CC}$

The monostable multivibrator will remain in its stable state until it receives an external trigger pulse which will cause it to change from its stable state to its semistable state. The monostable circuit remains in the semistable state for a time which is determined by the time constant RC.

Monostable Multivibrator Operation

The following description of the operation of the monostable multivibrator is based on the circuit shown in figure 7-4, page 51. Initially, the monostable circuit is in its stable state where Q1 is off and Q2 is on. The capacitor C will charge to the value V_{CC} and the output voltage at the collector of Q2 will be at the ground potential. This voltage level is fed to the base of Q1 and Q1 remains in the off state.

If a negative pulse of sufficient amplitude and duration is applied to the base of Q2, Q2 is reverse biased and turns off. As a result, the collector voltage of Q2 increases to the value V_{CC}. This increased Q2 collector voltage is coupled to the base of Q1 and Q1 is turned on. The Q1 collector voltage now goes to the ground potential and capacitor C begins to charge from a negative V_{CC} to a positive V_{CC}, figures 7-5 and 7-6. As the capacitor C charges toward $+V_{CC}$, a voltage level is reached which applies a forward bias

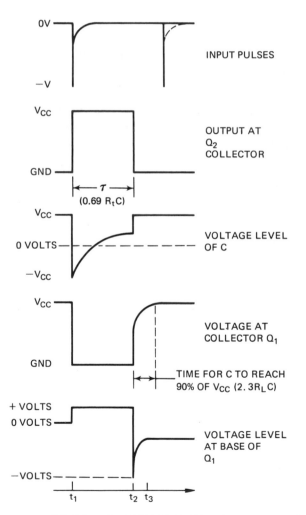

FIG. 7-7   TIMING DIAGRAM FOR MONO-
STABLE MULTIVIBRATOR

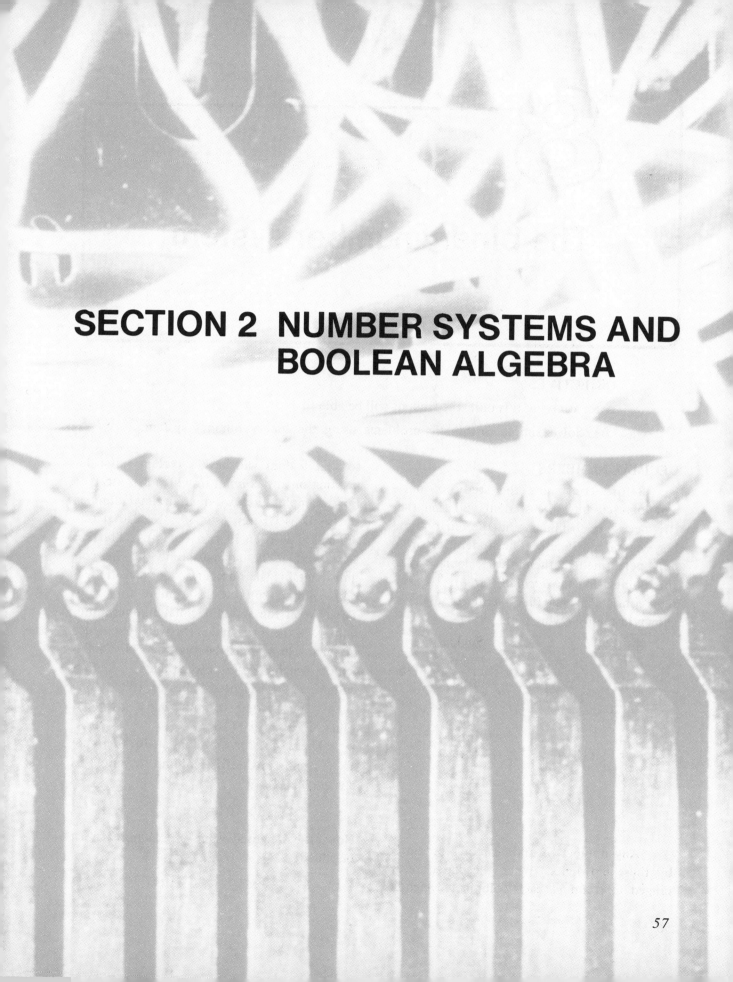

SECTION 2 NUMBER SYSTEMS AND BOOLEAN ALGEBRA

The binary number system

OBJECTIVES

After studying this unit, the student will be able to

- Solve simple arithmetic problems using the binary number system.

DECIMAL NUMBERS

Before continuing the study of pulse and logic circuits, the student will find that it is necessary to understand the binary number system and its use. The student has probably performed mathematical calculations to this point using the *decimal system*. This system consists of ten (deci) units or digits as shown in the following table.

Written	Spoken
0	zero
1	one
2	two
3	three
4	four
5	five
6	six
7	seven
8	eight
9	nine

Any number of units can be represented by these ten digits. When the digits are arranged in columns, each column is weighted so that a digit will have a certain value depending on the column in which it is located. For example, the two-digit decimal number thirty-six is written as: 36. Note that this number is written in two columns. Starting to the left of the decimal point, the first column contains the number six (6) and the second column contains the number three (3). To arrive at the actual decimal number represented by the written 36, the weighted values of each column must be expressed. Each column is weighted as a power of ten. The first column has a weighted value of 10^0 or one (1). The second column has a weighted value of 10^1 or ten (10).

Therefore, for the number 36, there are six (6) units in the first column, weighted to a value of 10^0:

$$6 \times 10^0 = 6 \text{ units}$$

Now, the weighted value of column 2 must be added:

$$3 \times 10^1 = 30 \text{ units}$$

Column one plus column two yields:

$$30 + 6 = 36$$

The number 274 is expressed in the decimal system as follows:

$$4 \text{ units at } 10^0 = 4 \times 10^0 = 4$$
$$\text{plus} \quad 7 \text{ units at } 10^1 = 7 \times 10^1 = 70$$
$$\text{plus} \quad 2 \text{ units at } 10^2 = 2 \times 10^2 = \underline{200}$$
$$274$$

Since we all use the decimal system from the time we begin to learn mathematics, this explanation of the system is given only as a reminder of the way in which a number is represented.

THE BINARY NUMBER SYSTEM

The binary number system contains only two (bi) units or digits.

Written	Spoken
0	zero
1	one

As a result, any quantities which are to be represented in the binary system must be written using only the two digits zero and one. For example, the binary number (one, one) is written as: 11 (*this is not the number eleven*)

How many units are represented by this number? Each column is given a weighted value (just as in the decimal system). For this system, however, the weighted values are based on powers of two rather than on powers of ten. Therefore, for the number 11, the first column to the left of the decimal point has a weighted value of:

$$2^0 \text{ or one } (1)$$

The second column has a weighted value of:

$$2^1 \text{ or two } (2)$$

As a result the number (one, one) in the binary system represents three units.

Therefore, in column one

$$1 \text{ unit at } 2^0 = 1 \times 2^0 = 1$$

plus, in column two

$$1 \text{ unit at } 2^1 = 1 \times 2^1 = 2$$

Adding: $1 + 2 = 3$ units.

Remember, number systems are used to denote how many of something. Eleven units in the decimal system is still eleven units in the binary system. However, the written representations of the number are different.

Spoken	Decimal (Written)	Binary (Written)
Eleven	11.	1011.
Twenty-Two	22.	10110.

Since the binary system consists of only two digits, it becomes quite awkward to write large numbers. For example, the number sixty-four is shown as a decimal number and as a binary number:

Decimal (Written)	Binary (Written)
64	1000000

WEIGHTING OF COLUMNS

Table 8-1, page 60, shows how the columns are weighted for both the binary number system and the decimal system.

For any number, to determine how many units are being represented, multiply the digit in any column by the weighting factor of that column and then add each answer.

This process is simple in the binary system since there are only two digits available, zero (0) and one (1).

PROBLEM 1

How many units are represented by the following binary numbers?

(a) 1010101

Solution

Place each of the digits in its appropriate column.

2^6	2^5	2^4	2^3	2^2	2^1	2^0
						. (radix)
1	0	1	0	1	0	1

Decimal														
10^n	. . .	10^4	10^3	10^2	10^1	10^0	.	10^{-1}	10^{-2}	10^{-3}	10^{-4}	. . . .	10^{-n}
						* (radix)							
Binary														
2^n	. . .	2^4	2^3	2^2	2^1	2^0	.	2^{-1}	2^{-2}	2^{-3}	2^{-4}	. . . .	2^{-n}
						* (radix)							

* The *radix* is the decimal point.

Table 8-1 Weighting of columns

The following operations are to be performed.

$1 \times 2^6 + 0 \times 2^5 + 1 \times 2^4 + 0 \times 2^3 + 1 \times 2^2 + 0 \times 2^1 + 1 \times 2^0$

The result is:

$1 \times 64 + 0 + 1 \times 16 + 0 + 1 \times 4 + 0 + 1 \times 1$

Adding these numbers yields:

$64 + 16 + 4 + 1 = 85$ units

(The final answer in this example, and in those that follow, is given in the decimal form simply because this system is the one which most people use most of the time.)

(b) 1101

Solution

Place each of the digits in the appropriate column.

2^3	2^2	2^1	2^0
1	1	0	1

Thus,

$1 \times 2^3 + 1 \times 2^2 + 0 \times 2^1 + 1 \times 2^0$

or $1 \times 8 + 1 \times 4 + 0 + 1 \times 1$

Adding, $8 + 4 + 1 = 13$ units

(c) 101.101

Solution

Place each of the digits in the appropriate column.

2^2	2^1	2^0	.	2^{-1}	2^{-2}	2^{-3}
1	0	1	.	1	0	1

Now break the problem into two halves. Use the decimal point (the radix) as the dividing point.

To the left of the radix:

$1 \times 2^2 + 0 \times 2^1 + 1 \times 2^0$

or, $1 \times 4 + 0 + 1 \times 1$

Adding, $4 + 1 = 5$ units

To the right of the radix:

$1 \times 2^{-1} + 0 \times 2^{-2} + 1 \times 2^{-3}$

$1 \times 0.5 + 0 \times 0.25 + 1 \times 0.125$

Adding, $0.5 + 0 + 0.125 = 0.625$ units

Combine the two halves to obtain the complete decimal number:

$5 + 0.625 =$

5.625 units

ADDITION OF BINARY NUMBERS

To add binary numbers, the following simple rules must be observed.

$1 + 0 =$ sum of 1 and carry of 0
$0 + 1 =$ sum of 1 and carry of 0
$1 + 1 =$ sum of 0 and carry of 1
$0 + 0 =$ sum of 0 and carry of 0

PROBLEM 2

Add binary 10 to binary 11 (decimal 2 to decimal 3).

2	10.
+3	11.
5	?

(a) The first column to the left of the radix contains a zero (0) and a one (1).

0
1

According to the binary addition rules, the sum of 0 and 1 is one (1) and a carry of zero (0).

```
    0        carry
    0
    1
    1
```

The carry is placed in the next column to the left.

```
  0 ◄─┐     carry
 ┌─────────┐
 │ 1    0  │ original
 │ 1    1  │ problem
 └─────────┘
      1
```

(b) The second column now contains three digits which are added two at a time.

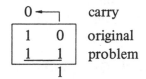

```
      ┌ 0   second
Add  ┤          column
      └ 1
        1    only
```

Adding the zero and one as shown gives a sum of *one* with a *carry* of zero.

```
carry─►0   sum─►1─┐ 0   second
                  │ 1   column only
carry─►1     ┌─►1─┘ 1   add this to one,
           carry       from one and
  ─────────────────    zero addition
              0
```

This one is added to the third one, giving a *zero* with a carry of *one*. The final one and zero (the carry digits) are now added together. The final answer is as follows:

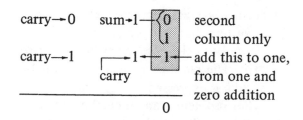

```
       ┌─►0 column 2       ┌─┐ ┌─┐
carry ─┤              │ 1│ │ 0│ column 1
       └─►1           │ 1│ │ 1│
          1           │ 0│ │ 1│
```

Add the following binary numbers.

```
 0 1 0
 1 0 1        (R8-1)
 ─────
 1 0 1
 1 1 0        (R8-2)
 ─────
 1 1 1
 1 0 1        (R8-3)
 ─────
```

SUBTRACTION OF BINARY NUMBERS

To subtract binary numbers, the set of rules listed below must be followed.

0 – 0 = difference of zero and borrow of zero
1 – 0 = difference of one and borrow of zero
0 – 1 = difference of one and borrow of one
1 – 1 = difference of zero and borrow of zero

PROBLEM 3

Subtract the binary 3 (011) from the binary five (101).

```
    1 0 1
  - 0 1 1
```

Start in the column of the least significant digit (at the right) and apply the subtraction rules given.

(a) One minus one gives a difference of zero and a borrow of zero. A borrow is always made from the column to the immediate left of the column in which the arithmetic operation is occurring.

```
borrowed   1 0 1
         - 0 1 1
               0
```

(b) Now move to the left to the next column: zero minus one. The difference of this column is one with a borrow of one. Since the borrow must be from the column to the immediate left, the third column is now blank. The one of the third column is "borrowed" to complete the arithmetic of the second column. The final answer for this subtraction problem is:

```
borrowed ─► ①0 1 = 5 ⎫ decimal
            0 1 1 = 3 ⎬ equivalent
            ─────────
              1 0 = 2 ⎭
```

Using the rules of binary arithmetic, solve the following problems in subtraction.

$$
\begin{array}{r}
1\ 1\ 1 \\
\underline{1\ 0\ 1} \\
\end{array}
\qquad (R8\text{-}4)
$$

$$
\begin{array}{r}
1\ 0\ 1 \\
\underline{0\ 0\ 1} \\
\end{array}
\qquad (R8\text{-}5)
$$

$$
\begin{array}{r}
1\ 1\ 0 \\
\underline{1\ 0\ 1} \\
\end{array}
\qquad (R8\text{-}6)
$$

SUBTRACTION BY THE COMPLEMENT METHOD

Binary numbers can also be subtracted by a method called the *complement method.*

PROBLEM 4

As in Problem 3, subtract the binary three (011) from the binary five (101).

(a) State the problem as shown.

$$
\begin{array}{r}
1\ 0\ 1 \\
-\ 0\ 1\ 1 \\
\end{array}
$$

(b) The next step is to complement the subtrahend. In other words, the value of each of the digits is reversed or complemented. The complement of one (1) is zero (0), and the complement of zero (0) is one (1). The problem is now written in the following form.

$$
\begin{array}{ll}
1\ 0\ 1 & \\
\underline{1\ 0\ 0} & \text{complement}
\end{array}
$$

(c) Instead of subtracting the two sets of numbers, they are now added.

$$
\begin{array}{r}
1\ 0\ 1 \\
\underline{1\ 0\ 0} \\
1\ 0\ 0\ 1
\end{array}
$$

(d) The final step is to take the most significant digit (MSD) of the answer (digit on the left) and add it to the least significant digit (LSD) (the digit on the right).

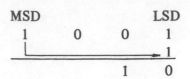

$$
\begin{array}{cc}
\text{MSD} & \text{LSD} \\
1 \quad 0 \quad 0 & 1 \\
\end{array}
$$

Thus, the final answer is 10, the binary expression for two.

The reason for the use of the subtraction by complement method will become apparent in the study of computer arithmetic operations in Unit 11.

CONVERSION OF DECIMAL NUMBERS TO BINARY NUMBERS

Earlier in this unit, a method was given for converting a binary number to its decimal equivalent. It is also important that the student be able to convert decimal numbers to binary numbers. The following example illustrates a method of performing this conversion.

PROBLEM 5

Convert the number seventy (70) from the decimal system (base ten system) to the binary system (base two system).

Seventy in the base ten system is written as 70_{10}. The required conversion in the base two system is expressed as X_2.

The problem can then be stated as: $70_{10} = X_2$. The method used to solve the problem is called the Quotient-Remainder method:

Divide the number to be converted by the base of the number system of the answer and record the remainder. Continue to divide the new quotients and record the remainders.

Therefore, to convert 70_{10} to its binary equivalent, first divide 70 by 2 (the base of the binary system).

$$2\,\overline{\smash{)}\,70}$$

The division yields a quotient of 35 and a remainder of 0.   Record this remainder. Now divide the new quotient (35) by 2 and so on.

$2\overline{)70}$
$2\overline{)35}$ + 0
$2\overline{)17}$ + 1
$2\overline{)8}$ + 1
$2\overline{)4}$ + 0
$2\overline{)2}$ + 0
$2\overline{)1}$ + 0
0   + 1

Write the remainders from left to right starting with the last remainder.

last remainder

1 0 0 0 1 1 0.

Therefore, 70_{10} = 1 0 0 0 1 1 0$_2$

Check the solution of Problem 5 by converting the binary number back to its decimal equivalent. (R8-7)

Convert 68_{10} *to its binary equivalent* $(68_{10} = X_2)$. *(R8-8)*

Complete the following chart.

	Binary	Decimal
(1)	0 0 0 1	1
(2)	———	2
(3)	0 0 1 1	——
(4)	0 1 0 0	4
(5)	———	5
(6)	0 1 1 0	——
(7)	0 1 1 1	7
(8)	———	8
(9)	———	9
(10)	———	10

(R8-9)

63

Boolean algebra

OBJECTIVES

After studying this unit, the student will be able to

- Apply the basic rules of Boolean algebra.
- Write a Boolean equation for a given set of conditions.

INTRODUCTION

In approximately 1847, George Boole wrote several papers on the mathematical analysis of logic. Boole's logic, which came to be known as *Boolean algebra,* is based on the assumption that a statement is either true or false. For this reason, Boolean algebra is sometimes called the *true - false logic.* Boolean algebra statements can be represented by the two digits of the binary number system. That is, the two numbers, 1 and 0, in the binary system are used to represent the true and false logic statements.

Both of these mathematical systems, the binary number system and Boolean algebra, are compatible with electronic circuits, as shown by the example in the following chart.

Electronic Circuit	Boolean Expression	Binary Digit
signal current or voltage	True	1
no signal current or voltage	False	0

A positive logic representation is shown in the chart. For a negative logic representation the reverse of the chart is used.

The AND and OR Symbols

One of the more confusing aspects of Boolean algebra for the beginner is the use of the symbols + and *x*. In Boolean algebra, these symbols *do not* mean add and multiply. The + symbol is assigned the name OR and the x symbol is assigned the name AND. The Boolean expression A + B is read as A OR B and the Boolean expression A x B is read as A AND B.

The student must know the following operations concerning the AND and OR terms before continuing the study of Boolean algebra.

1. 0 + 0 = 0 (zero OR zero equals zero)
2. 0 + 1 = 1 (zero OR one equals one)
3. 1 + 1 = 1 (one OR one equals one)
4. 0 x 0 = 0 (zero AND zero equal zero)
5. 0 x 1 = 0 (zero AND one equal zero)
6. 1 x 1 = 1 (one AND one equal one)

The following statements are true in Boolean algebra and are the rules by which Boolean algebra must be operated.

$$AB = A \times B = A \text{ AND } B$$
$$(A)(B) = A \times B = A \text{ AND } B$$
$$A \cdot B = A \times B = A \text{ AND } B$$

Also

$$A \times 1 = A \text{ (A and one equal A)}$$
$$A \times 0 = 0 \text{ (A and zero equal zero)}$$
$$A \times \overline{A} = 0 \text{ (A and not A equal zero)}$$
$$A \times A = A \text{ (A and A equal A)}$$

Similarly

$$A + 1 = 1 \text{ (A or one equals one)}$$
$$A + 0 = A \text{ (A or zero equals A)}$$
$$A + \overline{A} = A \text{ (A or not A equals A)}$$
$$A + A = A \text{ (A or A equals A)}$$
$$A(A + B) = A \text{ (A and A or B equal A)}$$
$$A + (AB) = A \text{ (A or A and B equals A)}$$

The final set of rules by which Boolean algebra operates consists of the Commutative Laws, the Associative Laws, and the Distributive Laws.

Commutive Laws

$$A \times B = B \times A \quad \text{multiplication}$$
$$A + B = B + A \quad \text{addition}$$

Associative Laws

$$A(B \times C) = (A \times B) C \quad \text{multiplication}$$
$$A + (B + C) = (A + B) + C \quad \text{addition}$$

Distributive Laws

$$A(B + C) = AB + AC \quad \text{multiplication}$$
$$A + (B \times C) = (A + B)(A + C) \quad \text{addition}$$
$$A + BC = (A + B)(A + C)$$

The truth of any of the Boolean algebra laws can be proved. For example, the steps which follow are a proof of the addition section of the Distributive Law.

Proof. $A + BC = (A + B)(A + C)$

1. The right-hand side of this expression can be expanded to:

$$(A + B)(A + C) = AA + AC + AB + BC$$

2. The common A can be factored from the first three terms of the expression to yield:

$$A(1 + C + B) + BC$$

3. Since A and anything equals A (from the rules applying to Boolean algebra), the A expression in step 2 becomes

$$A(1 + C + B) = A$$

4. Therefore

$$\mathbf{A + BC = A + BC}$$

To summarize:

$$\begin{aligned} A + BC &= (A + B)(A + C) \\ &= AA + AC + AB + BC \\ &= A(1 + C + B) + BC \\ &= A + BC \end{aligned}$$

USE OF BOOLEAN EQUATIONS

For the average technician, an extensive use of Boolean algebra will not be required. There will be times, however, when the rules of Boolean algebra can be applied to a circuit to help the technician understand the operation of the circuit.

The principles of Boolean algebra can be applied readily to the electrical circuits of the automobile.

In most automobiles, when the ignition is first turned on, prior to starting the engine, several lights on the dashboard turn on. In general, these lights are for the oil pressure indicator, the engine temperature indicator, and the battery charge/discharge indicator. Figure 9-1, page 66, illustrates what happens for the two conditions occurring when the ignition is on. When the ignition key is on, but the engine is not running, then these indicator lights are on. When the ignition key is on, and the engine is running and all of the systems are operating normally, then these indicators lamps are off.

To be effective, an indicator lamp must light when there is a problem with the automobile. For example, the following two

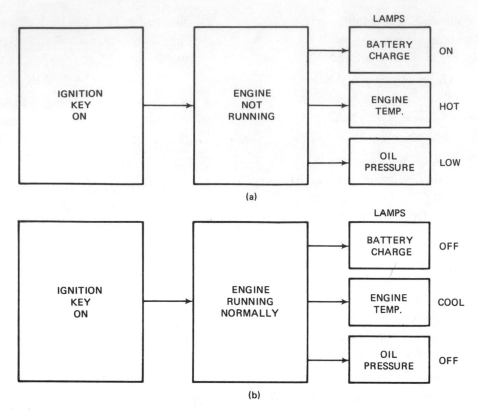

FIG. 9-1 BOOLEAN ALGEBRA ILLUSTRATION

conditions will cause the oil pressure lamp to light.

1. When the engine is not running and the ignition key is on.

2. When the engine is running but the oil pressure is low.

A Boolean equation can be written to specify the proper sequence of conditions for the operation of the oil pressure lamp. To write the equation, a code is assigned to each of the conditions encountered in the operation of the lamp.

Code	Name	Condition
A	A	ignition key on
$\overline{A}$	Not A	ignition key off
B	B	engine running
$\overline{B}$	Not B	engine not running
C	C	oil pressure at proper level
$\overline{C}$	Not C	oil pressure low

Two equations can be written for the two sets of conditions that cause the oil pressure indicator lamp to come on.

1. $(A) (\overline{B})$ = Lamp
2. $(A) (B) (\overline{C})$ = Lamp

Equations 1 and 2 can be combined to obtain:

$$A\overline{B} + AB\overline{C} = \text{Lamp}$$

This expression is read as:

A and not B or A and B and not C will complete the circuit to light the lamp.

RELAY CIRCUITS

As another example of the application of Boolean algebra, consider a relay circuit. Assume that a closed or short circuit is represented by a one (1), and an open circuit is represented by a zero (0). The relay symbols that will be used in relay circuit diagrams are shown in figure 9-2a, page 67.

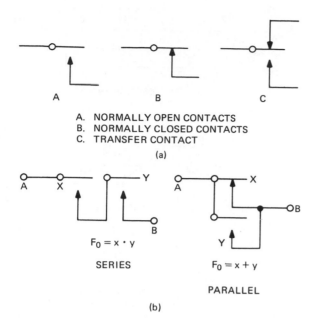

A. NORMALLY OPEN CONTACTS
B. NORMALLY CLOSED CONTACTS
C. TRANSFER CONTACT

(a)

$F_O = x \cdot y$

SERIES

$F_O = x + y$

PARALLEL

(b)

FIG. 9-2   RELAY SYMBOLS

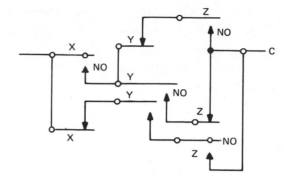

FIG. 9-3

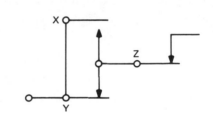

FIG. 9-4

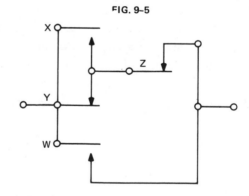

FIG. 9-5

In the series circuit of figure 9-2b, the transfer of a signal from point A to point B requires that both relays X and Y have their contacts closed at the same time. In other words, to obtain an output F_O, the following is true: (X) x (Y).

Therefore, the Boolean equation for the series circuit of figure 9-2b is $F_O = (X)$ Y or $F_O = X$ and Y.

In a similar manner, for the parallel circuit of figure 9-2b, either relay X or relay Y can be in the closed position and have a signal transferred from point A to point B. The Boolean equation for this circuit can be written as:

$$F_O = X + Y$$
$$(F_O = X \text{ or } Y)$$

PROBLEM 1

Design a two-terminal circuit using thee relays — X, Y, and Z — such that the circuit is closed when any two *but not three* of the relays are operated.

1. Write the Boolean expression of the conditions.

$$F(X, Y, Z) = XY\overline{Z} + X\overline{Y}Z + \overline{X}YZ$$

FIG. 9-6

2. Simplify the equation.

$$F(X, Y, Z) = X(Y\overline{Z} + \overline{Y}Z) + \overline{X}YZ$$

3. The circuit solution to the problem is shown in figure 9-3.

Write the Boolean expressions for the following circuits. (R9-1) Fig. 9-4 (R9-2) Fig. 9-5 (R9-3) Fig. 9-6

Several techniques have been developed to simplify the process of factoring the Boolean algebraic expression. One of the more useful techniques is the *Karnaugh map* shown in figure 9-7.

The following function can be factored or simplified by the use of the Karnaugh map.

$$F(X, Y, Z) = \overline{X}\,\overline{Y}\,Z + \overline{X}\,Y\,Z + X\,\overline{Y}\,\overline{Z} + X\,\overline{Y}\,Z$$

The value of zero is given to all bar terms and the value of one is given to all nonbar terms. It should be noted that the nonbar term X is read as X and the bar term $(\overline{X})$, X bar, is read as not X.

STEPS FOR THE KARNAUGH MAP

1. Prepare a chart as shown in figure 9-7. The columns from left to right are coded in the binary system (there must be a difference of only one binary digit between adjacent entries.

2. The X, Y, and Z values are placed in the boxes of the map by following a definite pattern, figure 9-8. Because there are three variables, each box will contain three quantities: X, Y, Z. The 0 and 1 values assigned to the left of and above the map indicate whether the variables are bar quantities or not. The 0 and 1 values to the left of the map designate the X variable, and the two numbers above each column designate the Y and Z variables respectively. Thus, if the first box of the map is filled in (X, Y, Z), the zeros to the left of and above the map indicate that each variable is a bar term; therefore, box one becomes $\overline{X}\,\overline{Y}\,\overline{Z}$. For the next box to the right, the zero to the left of the map indicates that the X is a bar term. The numbers above the second box indicate that the Y is a bar term and the Z has no bar: $\overline{X}\,\overline{Y}\,Z$. This procedure is followed until all of the map is complete.

3. Replace each of the complex values in each box of the map with a 1 (one) when the complex value corresponds to one of the

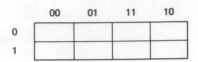

	00	01	11	10
0				
1				

FIG. 9-7   THE KARNAUGH MAP

	00	01	11	10
0	$\overline{X}\,\overline{Y}\,\overline{Z}$	$\overline{X}\,\overline{Y}\,Z$	$\overline{X}\,Y\,Z$	$\overline{X}\,Y\,\overline{Z}$
1	$X\,\overline{Y}\,\overline{Z}$	$X\,\overline{Y}\,Z$	XYZ	$XY\,\overline{Z}$

FIG. 9-8   STEP 2 FOR KARNAUGH MAPPING

	00	01	11	10
0	0	1	1	0
1	1	1	0	0

FIG. 9-9   STEP 3 FOR KARNAUGH MAPPING

terms in the original equation. Replace all other complex values with a 0 (zero), figure 9-9.

4. For the ones that are adjacent to each other in each row on the map, write the terms they represent as one equation which is to be factored. Therefore, for the first row, the terms represented by a one are: $\overline{X}\,\overline{Y}Z + \overline{X}\,Y\,Z$. This expression is factored as follows.

$$\begin{aligned} &\quad \overline{X}\,\overline{Y}\,Z + \overline{X}\,Y\,Z \\ =\;&\quad \overline{X}\,(\overline{Y}\,Z + Y\,Z) \\ =\;&\quad \overline{X} \cdot Z(\overline{Y}) + Y \\ =\;&\quad \overline{X}\,Z \end{aligned}$$

The expression for the second row is:

$$X\,\overline{Y}\,\overline{Z} + X\,\overline{Y}\,Z$$

Following a similar factoring procedure, the result for the second row expression is: $X\,\overline{Y}$ Therefore, the final answer is:

$$F(X, Y, Z) = \overline{X}\,\overline{Y}\,Z + \overline{X}\,Y\,Z + X\,\overline{Y}\,\overline{Z} + X\,\overline{Y}\,Z = \overline{X}\,Z + X\,\overline{Y}$$

PROBLEM 2

Factor the function $F(X, Y, Z) = \overline{X}\,\overline{Y}\,Z + \overline{X}\,Y\,\overline{Z} + \overline{X}\,Y\,Z = X\,\overline{Y}\,Z + X\,Y\,Z$

1. Draw the Karnaugh map as in figure 9-8. When the complex values present in the function are indicated on the map, figure 9-10 re-results. (Note that for three terms, the map has eight boxes since $2^3 = 8$.)

2. Group the adjacent one terms.

3. The middle group of four ones yields the term Z since Y and $\overline{Y}$ and X and $\overline{X}$ are present, $\overline{Z}$ is not present, and Z alone remains.

4. The group of two ones yields the term $\overline{X}$ Y (since Z and $\overline{Z}$ are present). Therefore, the function $F(X, Y, Z) = \overline{X} Y + Z$.

There are three additional points to be noted about the Karnaugh map. First, it is possible to group zeros instead of ones to obtain a negative of the function.

The second consideration to be noted is the *don't care condition*. For this condition, the output of the circuit is not important to the problem. This condition gives neither a zero nor a one on the Karnaugh map, but is

	00	01	11	10
0	0	1	1	1
1	0	1	1	0

FIG. 9-10   KARNAUGH MAP OF PROBLEM 2

given an X. This X can be grouped with either the ones or the zeros.

The third point concerns a shorthand way of writing the original function. For example, if a function has five variables as shown:

$$F(V, W, X, Y, Z) = \overline{V}\,\overline{W}\,\overline{X}\,\overline{Y}\,\overline{Z} + \overline{V}\,W\,\overline{X}\,Y\,Z + \ldots$$

the problem can be improved by writing the function in shorthand. Thus,

$$F(V, W, X, Y, Z) = 0, 1, 9, 11, 15, 16, 17, 24, 31$$

where each number represents a specific box of the map. A one is placed in every square that is represented by a number in the function, figure 9-11. For example, number 9 is square 01001 (binary 9) and represents the term $\overline{V} W \overline{X} \overline{Y} Z$.

		0	1	3	2	6	7	5	4
	XYZ								
	VW	000	001	011	010	110	111	101	100
0	00	1	1	0	0	0	0	0	0
8	01	0	1	1	0	0	1	0	0
24	11	1	0	0	0	0	1	0	0
16	10	1	1	0	0	0	0	0	0

FIG. 9-11   SIMPLIFIED KARNAUGH MAP REPRESENTATION

EXTENDED STUDY TOPICS

1. Draw a relay diagram for the automobile ignition problem presented in figure 9-1.

2. Write a Boolean equation for the following set of conditions. A signal lamp is to be lit when the following are true.

   a. I will go to the movie with a friend if it is Monday or Thursday.
   b. If there is a western or a comedy playing.
   c. If there is not a football game or sports special on TV.
   d. It is not snowing.
   e. I have the price of the admission to the theater.

SECTION 3 LOGIC

Unit 10

Logic gates

OBJECTIVES

After studying this unit, the student will be able to

- Design and construct circuits using logic gates.
- Construct logic circuits using integrated circuit components.
- Analyze logic circuits by means of truth tables.

DIODE LOGIC GATES

The diode may be regarded as an ideal electronic switch. When it is forward biased as shown in figure 10-1a, it acts as a very low resistance and allows current to flow easily. However, when it is reverse biased, figure 10-1b, it acts as a very high resistance and does not conduct. If two diodes are connected as shown in figure 10-2, and a small positive voltage is placed at points X and Y, the diodes will be forward biased and thus will conduct. The output, F (X, Y), takes on the value of the higher voltage, either X or Y.

Assume that the binary number 1 (one) represents a positive voltage which is sufficient to forward bias the diodes. A low voltage or ground is represented by the binary number 0 (zero). When a positive level of voltage (binary 1) is placed at either X or Y, there will be a voltage level of binary one at the output terminal. The circuit of figure 10-2

is called a two-input positive logic OR gate. In other words, the output of the circuit will be high (a logic 1) if either X *or* Y is high. This circuit will gate an input through to an output, in much the same way as

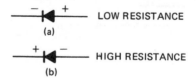

FIG. 10-1

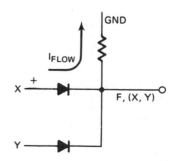

FIG. 10-2   SIMPLE DIODE "OR" GATE

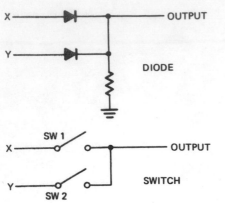

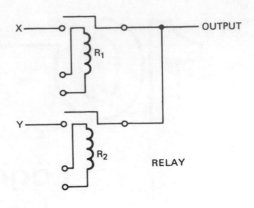

FIG. 10-3   OR GATES

a switch or set of relay contacts pass an input to an output, figure 10-3.

Another type of logic gate can be constructed using the same diodes with a slightly different bias condition, figure 10-4. As long as either input to this circuit (X or Y) is held at a low logic level (binary zero), the diode will be forward biased and the output will be at the same low logic level. Since a forward biased diode acts as a short circuit, the output

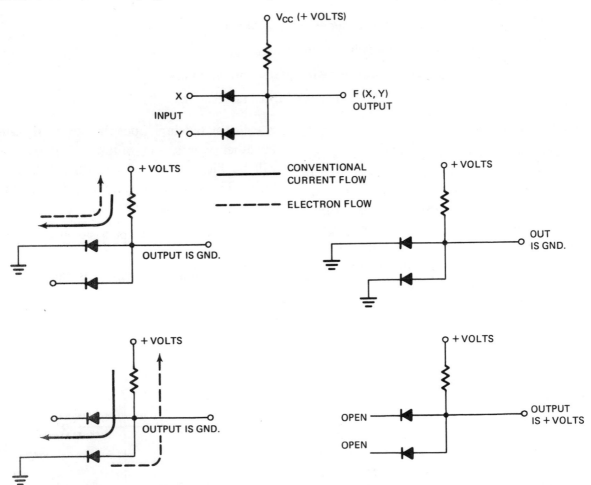

FIG. 10-4   THE DIODE "AND" GATE

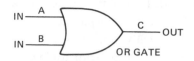

<table>
AND GATE
TRUTH TABLE

A	B	C
0	0	0
0	1	0
1	0	0
1	1	1
</table>

OR GATE
TRUTH TABLE

A	B	C
0	0	0
0	1	1
1	0	1
1	1	1

FIG. 10-5   LOGIC GATE SYMBOLS WITH TRUTH TABLES

will be shorted to a low logic level. However, if the X and Y inputs go to high at the same time, both diode circuits are opened and the voltage level felt at F(X, Y) is a logic level one.   Since both inputs must go high to achieve a logic level one at the output, this circuit is called a two-input positive logic *AND gate.*

Figure 10-5 shows the logic symbols for the AND gate circuit and the OR gate circuit and their *truth tables.*   Logic symbols only will be used in logic diagrams for the balance of this text.   Note that the logic symbol in-

dicates only the inputs and outputs.   The required grounds and supply voltages generally are not shown on logic diagrams.

Although figure 10-5 shows only gates with two inputs, gates with multiple inputs are available.   The operation of these gates is the same as that of the two-input gates shown.   For example, a three-input OR gate will have a logic level one output if a logic one appears at any of its three inputs.   A four-input AND gate must have a logic level one at each of its four inputs to produce a logic one level at its output.

LABORATORY EXERCISE 10-1: AND GATES

PURPOSE

- To verify the operation of AND gates

PROCEDURE

A.   Construct a two-input AND gate using two diodes and a resistor as shown in figure 10-4.
B.   Verify the truth table of figure 10-5 for a two-input AND gate.
C.   Add an additional diode to construct a three-input AND gate.
D.   Verify a truth table for this circuit.
     What step is necessary to obtain a logic level zero at any input?

TRANSISTOR LOGIC GATES

Diode logic gates were presented previously to help the student understand the principles of logic gate operation.   However,

at present, the most commonly used logic gates are *transistor logic gates.*   When studying transistor logic gates, the student will often see the abbreviations RTL, DTL, and TTL.

The last two letters in each of these abbreviations (TL) mean transistor logic. The first letter (R, D, or T) identifies the type of input used to feed the transistor logic gate.

Therefore,

RTL = resistor transistor logic (resistive input)

DTL = diode transistor logic (diode input)

TTL = transistor transistor logic (transistor input)

Of the three types of transistor logic gates, the TTL gate probably is the most commonly used.

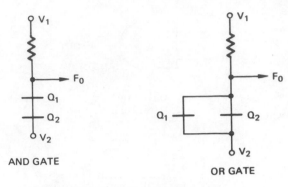

V_1 AND V_2 ARE VOLTAGE LEVELS
TRANSISTORS Q_1 AND Q_2 ARE SHOWN SYMBOLICALLY

FIG. 10-6

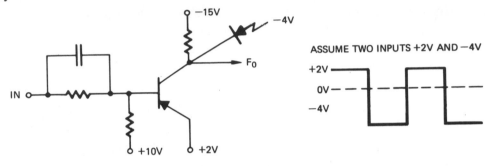

FIG. 10-7 PNP TRANSISTOR INVERTER

TRANSISTOR GATE OPERATION

If the proper conventions are observed, it can always be assumed that two transistors in a series switching arrangement form an AND gate; also, two transistors in parallel form an OR gate, figure 10-6.

Before continuing with the discussion of transistor gate operation, positive logic and negative logic must be defined.

Positive Logic

The more positive or high potential represents one (1) and a more negative or low potential represents zero (0).

Negative Logic

The more negative or low potential represents one (1) and a more positive or high potential represents zero (0).

The PNP Transistor Inverter

For the circuit in figure 10-7, -4 V applied at the input will forward bias the transistor so that it saturates (conducts) and the output is +2 V.

If +2 V is applied at the input, the transistor will be reverse biased or cut off. As a result, the output will be clamped at -4 volts.

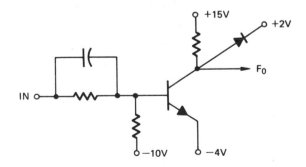

ASSUME TWO INPUTS; − 4V AND +2V

FIG. 10-8 NPN TRANSISTOR INVERTER

74

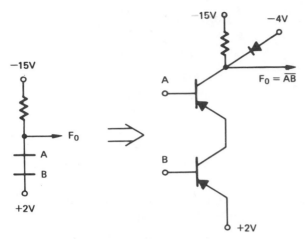

FIG. 10-9 PNP TRANSISTOR INVERTER AND GATE

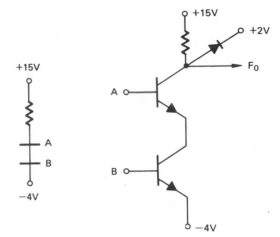

FIG. 10-10 NPN TRANSISTOR INVERTER AND GATE

Therefore, this circuit is classified as an inverter circuit.

The NPN Transistor Inverter

For the circuit in figure 10-8, -4 V applied at the input causes the transistor to be reverse biased or cut off. As a result, the output is clamped to +2 V. If +2 V is applied at the input, the transistor is forward biased or saturated and the output is -4 V. Therefore, this circuit is classified as an inverter.

The PNP Transistor Inverter AND gate

The circuit shown in figure 10-9 is a PNP series transistor AND gate.

Negative logic will be assumed for this circuit:

logic 0 = +2 V
logic 1 = -4 V

Therefore, $F_O = \overline{AB} + \overline{A}B + A\overline{B} = \overline{AB}$ Since it is assumed that two transistors in series form an AND gate, the functions ANDed together are inverted so that:

A and B in = $\overline{A \text{ and } B}$ out

A	B	F_O
0	0	1
0	1	1
1	0	1
1	1	0

Truth Table for PNP Series AND Gate

The NPN Transistor Inverter AND gate

The circuit shown in figure 10-10 is an NPN series AND gate.

Positive logic is assumed for this circuit:

logic 0 = -4 V
logic 1 = +2 V

Therefore, $F_O = \overline{AB} + A\overline{B} + \overline{A}B = \overline{AB}$ Since it was assumed that two transistors in series form an AND gate, the functions ANDed together are inverted:

$$\left[A \text{ and } B \text{ in}\right] = \left[\overline{A \cdot B} \text{ out}\right]$$

A	B	F_O
0	0	1
0	1	1
1	0	1
1	1	0

Truth Table for NPN Series AND Gate

The PNP Transistor Inverter OR gate

For the circuit shown in figure 10-11, page 76, the output is:

$$F_O = \overline{AB} = \overline{A + B}$$

Therefore, since it is assumed that two or more transistors in parallel form an OR gate, the functions ORed together are inverted:

$$\left[A \text{ or } B \text{ in}\right] = \left[\overline{A + B} \text{ out}\right]$$

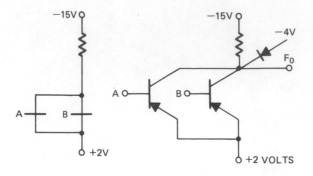

FIG. 10-11 PNP TRANSISTOR INVERTER OR GATE

A	B	F_O
0	0	1
0	1	0
1	0	0
1	1	0

Truth Table for PNP Inverter OR Gate

The NPN Transistor Inverter OR gate

The circuit shown in figure 10-12 is an NPN parallel OR gate. Positive logic is assumed for this circuit.

Therefore,

$$logic\ 0\ =\ -4\ V$$
$$logic\ 1\ =\ +2\ V$$
$$and\ F_O\ =\ \overline{AB}\ =\ \overline{A+B}$$

Since it is assumed that two or more transistors in parallel form an OR gate, the functions ORed together are inverted:

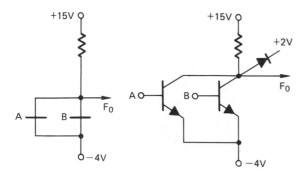

FIG. 10-12 NPN TRANSISTOR INVERTER OR GATE

$[A\ or\ B\ in]$	=	$[\overline{A+B}\ out]$

A	B	F_O
0	0	1
0	1	0
1	0	0
1	1	0

Truth Table for NPN Parallel OR Gate

A number of conclusions can be reached as a result of the study of the various logic gate circuits.

1. An NPN or PNP *series AND* or *parallel OR* gate may be constructed if the proper logic is used.

2. A particular function may be generated using these gates if the desired output function is inverted and the desired circuit is drawn from this equation or function.

PROBLEM 1

$$If\ F_O\ =\ A\ \overline{B}\ C$$
$$then\ F_O\ =\ \overline{A}+B+\overline{C}$$

Draw an OR gate for the function F_O. The OR gate is shown in figure 10-13.

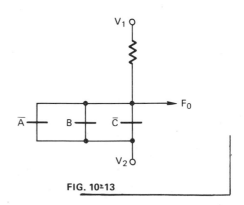

FIG. 10-13

PROBLEM 2

Construct a combination gate for the function:

$$F_O = A\bar{B} + \bar{A}BC$$
$$F_O = (\bar{A} + B)(A + \bar{B} + \bar{C}) = \bar{A}B + \bar{A}\bar{C} + AB + B\bar{C}$$

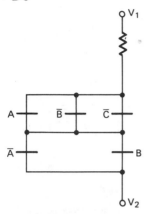

FIG. 10-14

Using the rules of logic gates, a gate can be constructed from any function.

NAND GATES

The previous section on transistor gate operation shows that transistor gate outputs are not the outputs shown in figure 10-5, but rather are these outputs inverted. It is for this reason that one of the most popular integrated circuit logic gates is the *NAND* gate, figure 10-15. Figure 10-15a shows the conventional symbol for a two-input AND gate. The output of this gate is then fed to an inverter circuit resulting in the inverse of the output. Figure 10-15b is the logic diagram symbol for a two-input positive logic NAND gate.

A commonly used NAND gate integrated circuit is the Texas Instrument type SN7400, figure 10-16, page 78. The integrated circuit package of the type 7400 device contains four individual gates. (These four gates give rise to the name *Quad* Two-input NAND Gate as shown on the specification sheet.) The circuit schematic diagram illustrates one advantage of the use of integrated circuit (IC) devices as compared to the use of discrete (single transistor and components) circuits. For the IC device, the inputs go to two emitters of the same transistor. This multiple-emitter technique is commonly used in integrated circuits. The student should recall that four identical circuits are contained in this single IC *chip* or package. Figure 10-15c shows the truth table for the NAND

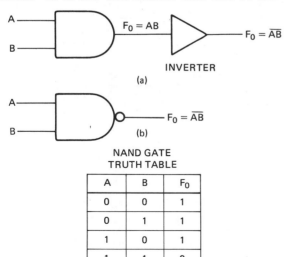

INVERTER

(a)

(b)

NAND GATE TRUTH TABLE

A	B	F_O
0	0	1
0	1	1
1	0	1
1	1	0

(c)

FIG. 10-15 NAND GATE WITH TRUTH TABLE

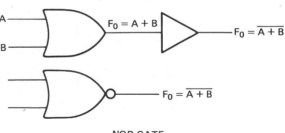

NOR GATE TRUTH TABLE

A	B	F_O
0	0	1
0	1	0
1	0	0
1	1	0

FIG. 10-17 NOR GATE WITH TRUTH TABLE

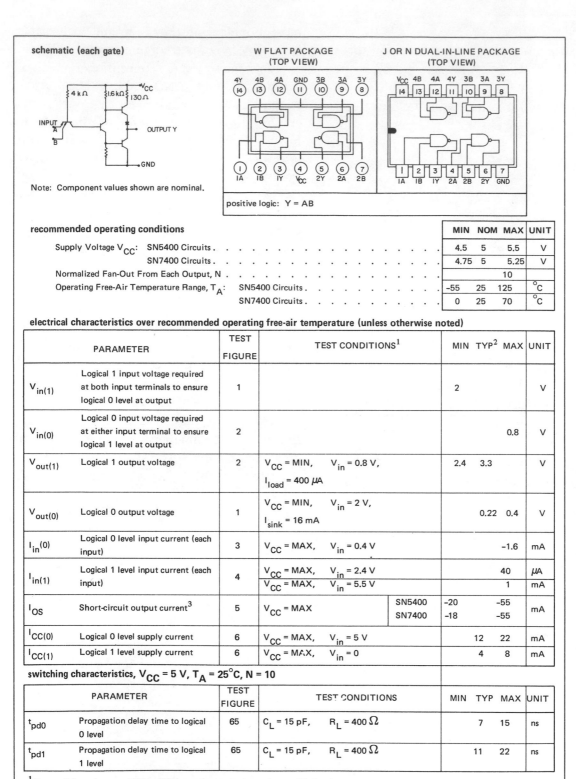

schematic (each gate)

Note: Component values shown are nominal.

W FLAT PACKAGE (TOP VIEW)

J OR N DUAL-IN-LINE PACKAGE (TOP VIEW)

positive logic: Y = AB

recommended operating conditions

		MIN	NOM	MAX	UNIT
Supply Voltage V_{CC}:	SN5400 Circuits	4.5	5	5.5	V
	SN7400 Circuits	4.75	5	5.25	V
Normalized Fan-Out From Each Output, N			10	
Operating Free-Air Temperature Range, T_A:	SN5400 Circuits	-55	25	125	$^{\circ}$C
	SN7400 Circuits	0	25	70	$^{\circ}$C

electrical characteristics over recommended operating free-air temperature (unless otherwise noted)

	PARAMETER	TEST FIGURE	TEST CONDITIONS[1]		MIN	TYP[2]	MAX	UNIT
$V_{in(1)}$	Logical 1 input voltage required at both input terminals to ensure logical 0 level at output	1			2			V
$V_{in(0)}$	Logical 0 input voltage required at either input terminal to ensure logical 1 level at output	2					0.8	V
$V_{out(1)}$	Logical 1 output voltage	2	V_{CC} = MIN, V_{in} = 0.8 V, I_{load} = 400 μA		2.4	3.3		V
$V_{out(0)}$	Logical 0 output voltage	1	V_{CC} = MIN, V_{in} = 2 V, I_{sink} = 16 mA			0.22	0.4	V
$I_{in(0)}$	Logical 0 level input current (each input)	3	V_{CC} = MAX, V_{in} = 0.4 V				-1.6	mA
$I_{in(1)}$	Logical 1 level input current (each input)	4	V_{CC} = MAX, V_{in} = 2.4 V				40	μA
			V_{CC} = MAX, V_{in} = 5.5 V				1	mA
I_{OS}	Short-circuit output current[3]	5	V_{CC} = MAX	SN5400	-20		-55	mA
				SN7400	-18		-55	
$I_{CC(0)}$	Logical 0 level supply current	6	V_{CC} = MAX, V_{in} = 5 V			12	22	mA
$I_{CC(1)}$	Logical 1 level supply current	6	V_{CC} = MAX, V_{in} = 0			4	8	mA

switching characteristics, V_{CC} = 5 V, T_A = 25°C, N = 10

	PARAMETER	TEST FIGURE	TEST CONDITIONS		MIN	TYP	MAX	UNIT
t_{pd0}	Propagation delay time to logical 0 level	65	C_L = 15 pF, R_L = 400 Ω			7	15	ns
t_{pd1}	Propagation delay time to logical 1 level	65	C_L = 15 pF, R_L = 400 Ω			11	22	ns

[1] For conditions shown as MIN or MAX, use the appropriate value specified under recommended operating conditions for the applicable device type.
[2] All typical values are at V_{CC} = 5 V, T_A = 25°C.
[3] Not more than one output should be shorted at a time.

FIG. 10-16 CIRCUIT TYPES SN5400, SN7400 QUADRUPLE 2-INPUT POSITIVE NAND GATES

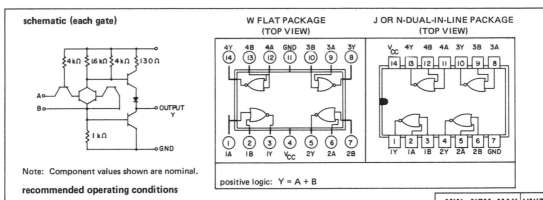

schematic (each gate)

Note: Component values shown are nominal.

recommended operating conditions

positive logic: Y = A + B

		MIN	NOM	MAX	UNIT
Supply Voltage V_{CC}:	SN5402 Circuits	4.5	5	5.5	V
	SN7402 Circuits	4.75	5	5.25	V
Normalized Fan-Out From Each Output, N				10	
Operating Free-Air Temperature Range, T_A:	SN5402 Circuits	-55	25	125	°C
	SN7402 Circuits	0	25	70	°C

electrical characteristics (over recommended operating free-air temperature range unless otherwise noted)

PARAMETER		TEST FIGURE	TEST CONDITIONS[1]		MIN	TYP[2]	MAX	UNIT
$V_{in(1)}$	Logical 1 input voltage required at either input terminal to ensure logical 0 level at output	8			2			V
$V_{in(0)}$	Logical 0 input voltage required at both input terminals to ensure logical 1 level at output	9					0.8	V
$V_{out(1)}$	Logical 1 output voltage	9	V_{CC} = MIN, V_{in} = 0.8 V, I_{load} = −400 μA		2.4	3.3		V
$V_{out(0)}$	Logical 0 output voltage	10	V_{CC} = MIN, V_{in} = 2 V, I_{sink} = 16 mA			0.22	0.4	V
$I_{in(0)}$	Logical 0 level input current (each input)	11	V_{CC} = MAX, V_{in} = 0.4 V				−1.6	mA
$I_{in(1)}$	Logical 1 level input current (each input)	12	V_{CC} = MAX, V_{in} = 2.4 V				40	μA
			V_{CC} = MAX, V_{in} = 5.5 V				1	mA
I_{OS}	Short-circuit output current[3]	13	V_{CC} = MAX	SN5402	−20		−55	mA
				SN7402	−18		−55	
$I_{CC(0)}$	Logical 0 level supply current	14	V_{CC} = MAX, V_{in} = 5 V			14	27	mA
$I_{CC(1)}$	Logical 1 level supply current	14	V_{CC} = MAX, V_{in} = 0			8	16	mA

switching characteristics, V_{CC} = 5 V, T_A = 25°C, N = 10

PARAMETER		TEST FIGURE	TEST CONDITIONS		MIN	TYP	MAX	UNIT
t_{pd0}	Propagation delay time to logical 0 level	65	C_L = 15 pF,	R_L = 400 Ω		8	15	ns
t_{pd1}	Propagation delay time to logical 1 level	65	C_L = 15 pF,	R_L = 400 Ω		12	22	ns

[1] For conditions shown as MIN or MAX, use the appropriate value specified under recommended operating conditions for the applicable device type.
[2] All typical values are at V_{CC} = 5 V, T_A = 25°C.
[3] Not more than one output should be shorted at a time.

FIG. 10-18 CIRCUIT TYPES SN5402, SN7402 QUADRUPLE 2-INPUT POSITIVE NOR GATES

EXCLUSIVE OR
TRUTH TABLE

A	B	F_0
0	0	0
0	1	1
1	0	1
1	1	0

FIG. 10–19   EXCLUSIVE OR GATE

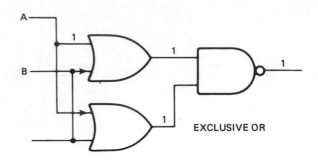

EXCLUSIVE OR

FIG. 10–20   CONSTRUCTING AN EXCLUSIVE OR GATE

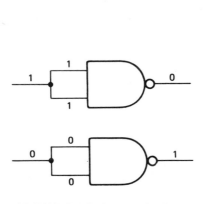

(a)   NAND GATE AS AN INVERTER

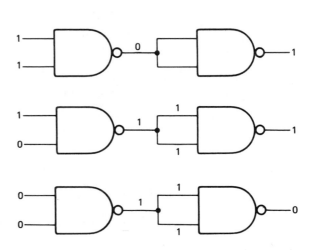

(b)   CONNECTING NANDS TO CONSTRUCT AN AND

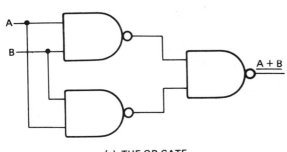

(c)   THE OR GATE

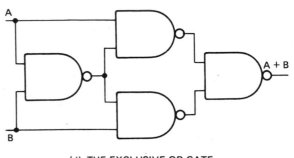

(d)   THE EXCLUSIVE OR GATE

FIG. 10–21   USING NAND GATES TO CONSTRUCT OTHER TYPES OF GATE CIRCUITS

gate. The two diodes at the input insure that large negative voltages at the input will not overdrive the transistor. Each of the four individual circuits in the IC package is tied to a common V_{CC} line and a common ground.

NOR GATES

Just as the NAND gate is a simple AND gate whose output is fed to an inverter circuit, the NOR gate is an OR gate plus an inverter, figure 10-17, page 77. A common integrated circuit package for the NOR gate is the Texas Instrument type SN7402, figure 10-18, page 79.

EXCLUSIVE-OR GATE

The logic symbol and the truth table for the EXCLUSIVE-OR gate are shown in figure 10-19. The truth table indicates that the EX-CLUSIVE-OR gate will have an output when either input A or input B is high. Unlike the conventional OR gate, if both input A and input B are high, the output will be low. In other words, the EXCLUSIVE-OR gate circuit operates *exclusively* in the OR state and not in the AND state. The logic block diagram in figure 10-20 shows the construction of an EXCLUSIVE-OR circuit using NAND and OR gates.

USING NAND GATES

The NAND gate is a very useful device. A number of other gates can be constructed using two or more NAND gates, figure 10-21. For example, figure 10-21a shows that if both the inputs are tied together, the NAND gate becomes an inverter circuit. In figure 10-21b, two NAND gates are connected to form an AND gate.

LABORATORY EXERCISE 10-2: THE NAND GATE

PURPOSE

- To investigate the properties and uses of the NAND gate.

PROCEDURE

Part I

A. Plug a type SN7400 IC NAND gate into a test socket and connect the ground and V_{CC} to the proper pins, figure 10-22.

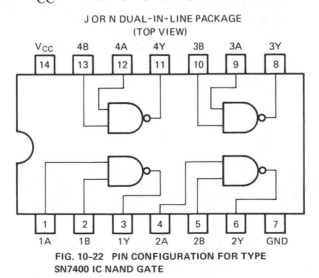

J OR N DUAL-IN-LINE PACKAGE
(TOP VIEW)

FIG. 10-22 PIN CONFIGURATION FOR TYPE
SN7400 IC NAND GATE

B. Check each of the four NAND gates of the IC for proper functioning by applying zero and one inputs and verifying the truth table. (Refer to figures 10-15, page 77 and 10-16, page 78.)

C. The specification sheet of figure 10-16, page 78, shows the discrete component schematic of the NAND gate. Note that the input leads are actually the emitters of a transistor. This is the equivalent of a logic one at that terminal. A logic zero at an input terminal is obtained by grounding the terminal.

D. To check the logic level of the output of each gate, connect a voltmeter or an LED (light-emitting diode) to the output as shown in figure 10-23.

Part II

A. 1. Once it is verified that each of the four NAND gates is functioning properly, construct the circuit of figure 10-24. The truth table for this circuit is as follows:

A	B	C
0	0	0
0	1	0
1	0	0
1	1	1

2. Apply the proper inputs and verify the truth table.

3. When a NAND gate serves as the input to another NAND gate, the resulting circuit should be an AND gate. A one, one input to gate 1 yields a zero output. Therefore, the input to gate 2 is zero, zero (ground is zero) and the output of gate 2 is one.

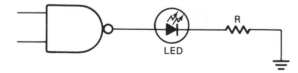

(a) LED IS LIGHTED FOR LOGIC LEVEL 1

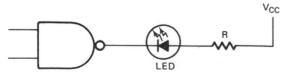

(b) LED IS LIGHTED FOR LOGIC LEVEL 0

FIG. 10-23

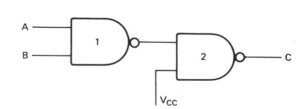

FIG. 10-24

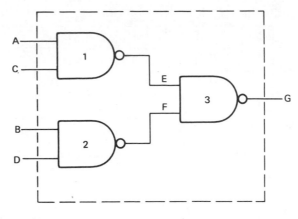

FIG. 10–25

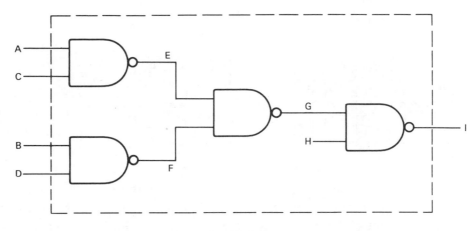

FIG. 10-26

Part III

A. Using three of the four gates of the SN7400 IC NAND gate, connect the circuit shown in figure 10-25.

B. Verify the following truth tables.

A	B	C	D	E	F	G
0	0	1	1	1	1	0
0	1	1	1	1	0	1
1	0	1	1	0	1	1
1	1	1	1	0	0	1

A	B	G
0	0	0
0	1	1
1	0	1
1	1	1

An examination of these truth tables will show that the circuit is an OR gate.

Part IV

A. Figure 10-26 shows another circuit that can be constructed using the NAND gate. Terminals C, D and H are tied to + volts providing a constant one input to these terminals.

B. Verify the following truth tables.

A	B	C	D	E	F	G	H	I
0	0	1	1	1	1	0	1	1
0	1	1	1	1	0	1	1	0
1	0	1	1	0	1	1	1	0
1	1	1	1	0	0	1	1	0

A	B	I
0	0	1
0	1	0
1	0	0
1	1	0

An examination of these truth tables will show that the circuit is a NOR gate.

Part V

A. A circuit using all four NAND gates of the type SN7400 IC is shown in figure 10-27.

B. Construct the circuit of figure 10-27.

C. Verify the following truth tables.

A	B	C	D	E
0	0	1	1	0
0	1	1	0	1
1	0	0	1	1
1	1	1	1	0

A	B	E
0	0	0
0	1	1
1	0	1
1	1	0

D. Compare the output of this circuit with the outputs of the other circuits tested. The circuit in figure 10-27 gives a one output when either the A input or the B input is a one, but not when both inputs are ones. This type of circuit is called an EXCLUSIVE-OR gate.

Is it important to know how each individual logic gate is constructed? Why? (R10-1)

Should a zero input be obtained by leaving the input terminal of a logic device disconnected? Why? (R10-2)

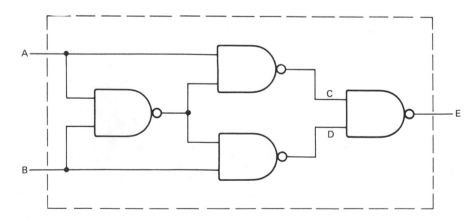

FIG. 10-27

How is a NAND gate connected to obtain an inverter circuit? (R10-3)
Draw a logic schematic diagram showing how one or more SN7400
ICs should be connected to construct a three-input AND gate. (R10-4)

REVIEW OF LOGIC LEVELS

The concept of logic levels as presented in this unit is very important and is summarized in figures 10-28 and 10-29.

Figure 10-28 shows a rectangular pulse with two levels which may be called the HI and LO levels. Figure 10-29 shows several ways of designating the HI and LO levels according to positive and negative logic. On many occasions, pulse inputs are assigned the one (1) and zero (0) values of the binary num-

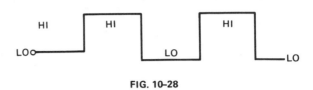

FIG. 10-28

ber system. In positive logic, the HI level is assigned a one (1) and the LO level is assigned a zero (0). In negative logic, the LO is assigned a one (1) and the HI is assigned a zero (0).

In figure 10-29, which of the pulses shown exhibit positive logic? (R10-5)

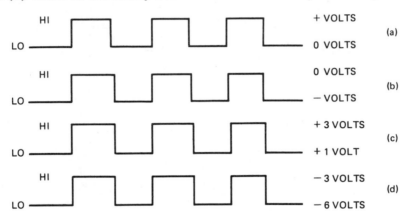

FIG. 10-29 POSITIVE AND NEGATIVE LOGIC

EXTENDED STUDY TOPICS

1. For the circuit of figure 10-30, write the Boolean expression for the output F_O. Using NAND gates only, construct a circuit to achieve this output.

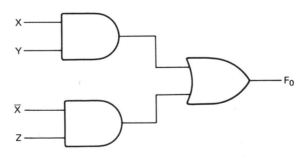

FIG. 10-30

2. Design the logic circuit for the Boolean equation

$$(A + B + C)(D + E) + F + G(H + I)J = K$$

3. Write the Boolean equation for the circuit of figure 10-31.

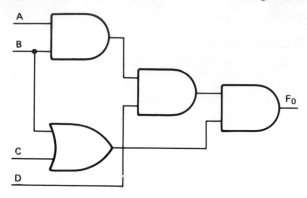

FIG. 10–31

Arithmetic logic gates

OBJECTIVES

After studying this unit, the student will be able to

- Construct a half adder and a full adder using NAND gates.
- Verify the operation of an integrated circuit full adder on one IC chip.

COMPUTER BUILDING BLOCKS

The previous units in this text investigated a number of individual logic circuits and concepts. This final unit will combine several of these circuits and concepts to form a small system which is used as one of the basic building blocks of a computer. Figure 11-1 shows a block diagram for a typical digital computer.

The section of the digital computer block diagram that will be covered in this unit is the arithmetic logic unit. It is within this unit that the computer performs all of its arithmetic operations. Most of these operations involve either binary addition or binary subtraction.

THE ADDER CIRCUIT

The requirements of an adder circuit for a digital computer can be illustrated by examining the simplest addition procedure — the addition of two binary digits. Table 11-1,

page 88, shows the logical requirements of a two-digit binary adder circuit. Boolean equations for the sum and carry terms can be written as follows.

$$\text{Sum} = X\overline{Y} + \overline{X}Y$$
$$\text{Carry} = XY$$

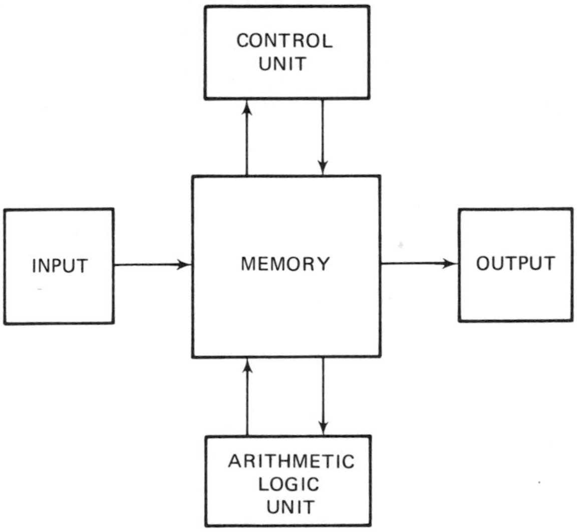

FIG. 11-1   BLOCK DIAGRAM FOR A DIGITAL COMPUTER

AUGEND	X	0	1	0	1
ADDEND	Y	0	0	1	1
SUM	S	0	1	1	0
CARRY	C	0	0	0	1

Table 11-1

The equation for the sum term of the binary addition is the same as the equation for an EXCLUSIVE-OR logic gate circuit (Unit 10). In addition, the equation for the carry term is the same as the equation for a logic AND gate. *Write the truth table for an EXCLUSIVE-OR gate (R11-1)*

It can be seen that the circuits required to perform the addition of two binary numbers are an AND gate and an EXCLUSIVE-OR gate. Figure 11-2 shows the logic for the adder circuit of the computer arithmetic logic unit.

Can the circuit of figure 11-2 be constructed using NAND gates? (R11-2)

For the circuit in figure 11-2, write the logic levels that can be expected at points A, B, sum, and carry for the addition of X = 1, Y = 0. (R11-3)

If the logic level at the X input of figure 11-2 is one, then it is expected that there will be a zero (0) at the carry terminal since the inputs to this gate are a logic one (1) and a logic zero (0). AND gate I has a one (1) and one (1) at its inputs due to the fact that the Y input is inverted before going to gate I. Therefore, point A of figure 11-2 is at a logic level one (1). By the same reasoning, AND gate II has a zero (0), zero (0) at its inputs. Thus, the logic level at point B is zero (0). Points A and B are inputs to the OR gate. The one (1) and zero (0) at this gate produce a logic level one (1) at the sum terminal of the circuit.

Do the logic levels determined for the circuit of figure 11-2 agree with Table 11-1? (R11-4)

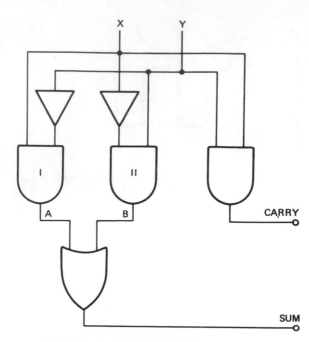

FIG. 11-2 LOGIC CIRCUIT FOR A HALF ADDER

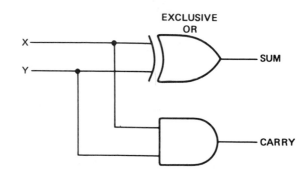

FIG. 11-3 SIMPLIFIED HALF ADDER LOGIC DIAGRAM

The circuit of figure 11-2 is called a *half adder* and is summarized in figure 11-3. Although the half adder circuit can adequately perform the addition of two binary numbers, it cannot accept a *carry input* from a preceding stage.

Is the half adder circuit limited as a binary adder? (R11-5)

THE FULL ADDER CIRCUIT

The capacity of the half adder circuit can be increased to enable it to perform the

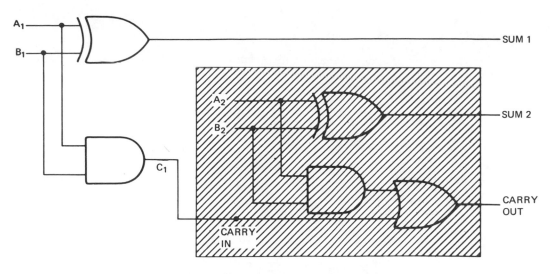

FIG. 11-4   THE FULL ADDER LOGIC DIAGRAM

arithmetic addition of problems such as the following one.

Add   1   1 (binary 3)
        0   1 (binary 1)

Solution C_1

	1	1		A_2	A_1
	0	1		B_2	B_1
1	0	0	C_O	S_2	S_1

To solve this problem, first assign to the binary number three (11) the A term ($A_2 A_1$). In addition, the binary number one (01) is assigned the B term ($B_2 B_1$). When A_1 is added to B_1, the result consists of terms S_1 (the sum) and C_1 (the carry). This carry must be added to the next column of digits (A_2, B_2). The addition of $A_1 B_1$ can be accomplished by the half adder circuit. The addition of $A_2 B_2$ and C_1, however, requires a circuit capable of taking three inputs and giving an output of a sum and carry, figure 11-4. Figure 11-4 is the logic diagram of a full adder circuit. Note that the section outside of the shaded portion is the half adder circuit. The circuit in the shaded area is the combination of a half adder circuit and an OR gate (the shaded area is actually the full adder).

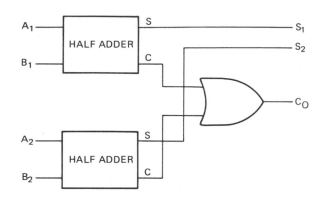

FIG. 11-5   FULL ADDER BLOCK DIAGRAM

Figure 11-5 is the block diagram of a full adder circuit.

What is the major difference between the full adder and the half adder? (R11-6)

Can the half adder circuit give a sum and carry out for all possible input conditions? (R11-7)

BINARY SUBTRACTERS

A logical subtract circuit can be developed in a manner similar to that for the logical adder circuit. Using Table 11-2, page 90, equations can be written for the difference (D) and borrow (B) factors in subtraction.

Minuend	X	0	1	0	1
Subtrahend	Y	0	0	1	1
Difference	D	0	1	1	0
Borrow	B	0	0	1	0

Table 11-2

The equations for a subtracter circuit are:

$$D = X\overline{Y} + \overline{X}Y$$
$$B = \overline{X}Y$$

These equations yield the equivalent of a half subtracter circuit due to the fact that no provision was made for a borrow from a previous stage.

A *full subtracter* circuit provides for this borrow feature.

Minuend	X	0	1	0	1	0	1	0	1
Subtrahend	Y	0	0	1	1	0	0	1	1
Borrow₁	B_{in}	0	0	0	0	1	1	1	1
Difference	D	0	1	1	0	1	0	0	1
Borrow₂	B_{out}	0	0	1	0	1	0	1	1

Table 11-3

Using Table 11-3, the following difference (D) and borrow (B) equations can be written.

$$D = X\overline{Y}\overline{B} + \overline{X}Y\overline{B} + \overline{X}\,\overline{Y}B + XYB$$
$$B = \overline{X}Y\overline{B} + \overline{X}\,\overline{Y}B + \overline{X}YB + XYB$$

Figure 11-6 is a logic block diagram for the full subtracter circuit.

INTEGRATED CIRCUIT FULL ADDERS

The circuits presented in this unit thus far are useful for helping the student to understand the operation of binary adder circuits. However, these circuits are not too practical. The small pocket calculators on the market today would be much larger if their circuits were similar to those described in this unit. The adder circuits used in these calculators are complete and are all contained on a single IC chip.

The operation of a full adder IC device can be analyzed by examining a typical device such as the Texas Instrument type SN7480 gated full adder integrated circuit. Figure 11-7 is

FIG. 11-7 SIMPLIFIED BLOCK DIAGRAM OF THE SN7480 GATED FULL ADDER

the block diagram for the SN7480. To use this device in a situation calling for addition of a binary one to a binary one with a binary one carry from a preceding stage, the following conditions must be met.

1. The following pins are grounded: pins 7, 12, 10 and 11.
2. Apply V_{CC} to pin 14
3. Apply the A input to pins 8 and 9
4. Apply the B input to pins 12 and 13

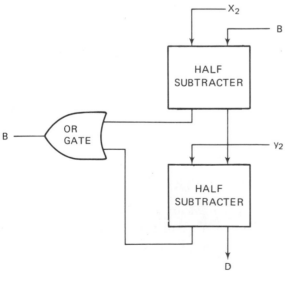

FIG. 11-6 FULL SUBTRACTER

5. For a carry out, (C_O), pin 4 must be fed to an inverter.

6. The truth table for this device is given in Table 11-4.

C_{IN}	B	A	$\overline{C_{OUT}}$	$\overline{S}$	S
0	0	0	1	1	0
0	0	1	1	0	1
0	1	0	1	0	1
0	1	1	0	1	0
1	0	0	1	0	1
1	0	1	0	1	0
1	1	0	0	1	0
1	1	1	0	0	1

TABLE 11-4 TRUTH TABLE FOR SN7480 GATED FULL ADDER

Figure 11-8 shows the wiring required if the SN7480 IC is to accomplish the desired addition. The battery in figure 11-8 should have a value of approximately 4.8 volts. This value will supply the proper value of V_{CC} to the device. In addition, this value of 4.8 volts insures that a logic one value is applied to the inputs of the logic gate. The inverter circuit between pin 4 and the carry out terminal is necessary since the output from pin 4 is the logic term $\overline{C_O}$ (not carry out). According to the truth table for the SN7480 full adder, the outputs should be as shown in the truth table of figure 11-9.

Figure 11-10, page 92, is a portion of a specification sheet for the SN7480 device from a TI integrated circuit handbook. This specification shows the logic diagram, truth table, circuit schematic, and pertinent information about the device. It can be seen that the logic diagram for this device is much more complex than any that have been examined thus far in this text. However, the block diagram and truth table in figures 11-7 and 11-8 provide sufficient information for the use of the device in a number of common situations.

SUBTRACTION BY ADDITION

The SN7480 device can be used as a subtracter as well as an adder. To subtract one number from another, the complement method can be used. In other words, the subtrahend is complemented and this value is then added to the minuend. Therefore, it is possible to perform subtraction by using an adder circuit and complementing one of the inputs of the circuit. The construction of the arithmetic unit is less complex if this method is used since the unit will require only an adder and several inverters.

OTHER TYPES OF IC ADDERS

Figures 11-11, page 94, and 11-12, page 95, show two additional adder circuits. Note in figure 11-11 the complex truth table for the type SN7482 2-bit full adder. Figure 11-12 shows the logic circuitry and truth table for a 4-bit full adder. This device is used to accomplish the logical addition of two four-bit binary numbers. This device will give four sum bit outputs and a carry out. Although the

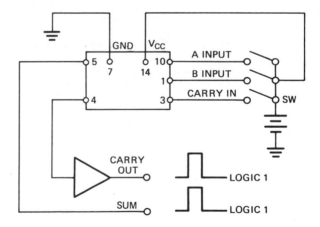

FIG. 11-8 CIRCUIT TO ADD A = 1, B = 1, C_{IN} = 1

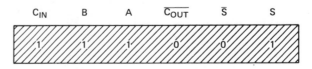

C_{IN}	B	A	$\overline{C_{OUT}}$	$\overline{S}$	S
1	1	1	0	0	1

FIG. 11-9 TRUTH TABLE FOR SN7480 SHOWS THE LOGIC LEVELS FOR FIG. 11-8

logic

FUNCTION TABLE
(See Notes 1, 2, and 3)

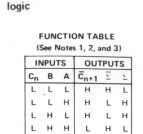

INPUTS			OUTPUTS		
C_n	B	A	$\overline{C}_{n+1}$	$\overline{\Sigma}$	Σ
L	L	L	H	H	L
L	L	H	H	L	H
L	H	L	H	L	H
L	H	H	L	H	L
H	L	L	H	L	H
H	L	H	L	H	L
H	H	L	L	H	L
H	H	H	L	L	H

H = high level, L = low level

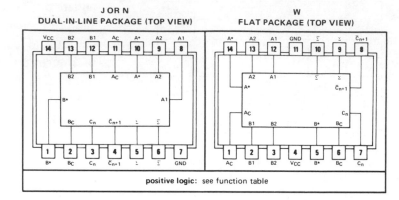

J OR N
DUAL-IN-LINE PACKAGE (TOP VIEW)

W
FLAT PACKAGE (TOP VIEW)

positive logic: see function table

NOTES: 1. $A = \overline{A}_C \cdot \overline{A*} + A1 \cdot A2$, $B = \overline{B}_C + \overline{B*} + B1 \cdot B2$.
2. When A* is used as an input, A1 or A2 must be low. When B* is used as an input, B1 or B2 must be low.
3. When A1 and A2 or B1 and B2 are used as inputs, A* or B*, respectively, must be open or used to perform dot-AND logic.

description

These single-bit, high-speed, binary full adders with gated complementary inputs, complementary sum (Σ and $\overline{\Sigma}$) outputs and inverted carry output are designed for medium- and high-speed, multiple-bit, parallel-add/serial-carry applications. These circuits (see schematic) utilize diode-transistor logic (DTL) for the gated inputs, and high-speed, high-fan-out transistor-transistor logic (TTL) for the sum and carry outputs and are entirely compatible with both DTL and TTL logic families. The implementation of a single-inversion, high-speed, Darlington-connected serial-carry circuit minimizes the necessity for extensive "look-ahead" and carry-cascading circuits.

absolute maximum ratings over operating free-air temperature range (unless otherwise noted)

Supply voltage, V_{CC} (see Note 4) . . . . 7 V
Input voltage (see Note 5) . . . . 5.5 V
Operating free-air temperature range: SN5480 Circuits . . . . −55°C to 125°C
SN7480 Circuits . . . . 0° to 70°C
Storage temperature range . . . . −65°C to 150°C

NOTES: 4. Voltage values are with respect to network ground terminal.
5. Input signals must be zero or positive with respect to network ground terminal.

recommended operating conditions

		SN5480			SN7480			UNIT
		MIN	NOM	MAX	MIN	NOM	MAX	
Supply voltage, V_{CC}		4.5	5	5.5	4.75	5	5.25	V
High-level output current, I_{OH}	Σ or $\overline{\Sigma}$			−400			−400	µA
	$\overline{C}_{n+1}$			−200			−200	
	A* or B*			−120			−120	
Low-level output current, I_{OL}	Σ or $\overline{\Sigma}$			16			16	mA
	$\overline{C}_{n+1}$			8			8	
	A* or B*			4.8			4.8	
Operating free-air temperature, T_A		−55		125	0		70	°C

FIG. 11-10 TYPES SN5480, SN7480 GATED FULL ADDERS (Continued)

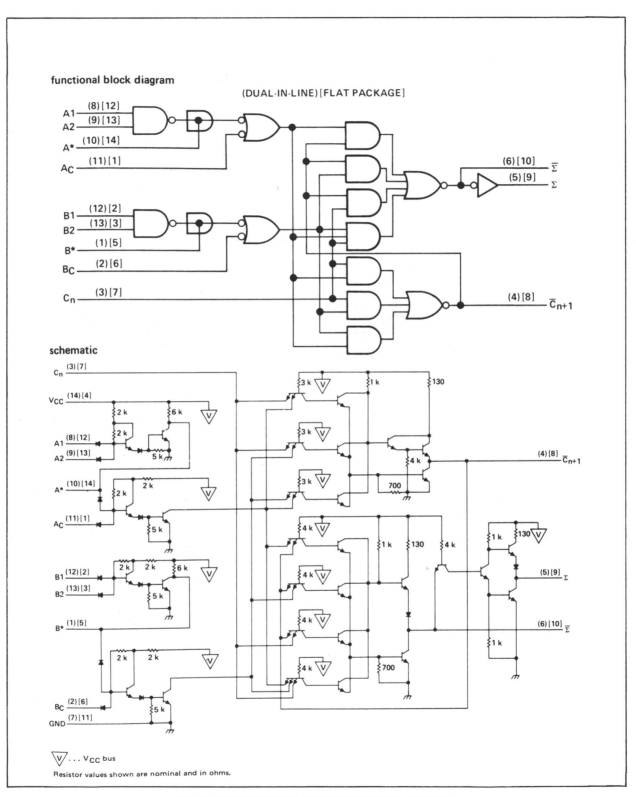

FIG. 11-10 TYPES SN5480, SN7480 GATED FULL ADDERS

For applications in:

- Digital Computer Systems
- Data-Handling Systems
- Control Systems

logic

FUNCTION TABLE

INPUTS				OUTPUTS					
				WHEN C0 = L			WHEN C0 = H		
A1	B1	A2	B2	Σ1	Σ2	C2	Σ1	Σ2	C2
L	L	L	L	L	L	L	H	L	L
H	L	L	L	H	L	L	L	H	L
L	H	L	L	H	L	L	L	H	L
H	H	L	L	L	H	L	H	H	L
L	L	H	L	L	H	L	H	H	L
H	L	H	L	H	H	L	L	L	H
L	H	H	L	H	H	L	L	L	H
H	H	H	L	L	L	H	H	L	H
L	L	L	H	L	H	L	H	H	L
H	L	L	H	H	H	L	L	L	H
L	H	L	H	H	H	L	L	L	H
H	H	L	H	L	L	H	H	L	H
L	L	H	H	L	L	H	H	L	H
H	L	H	H	H	L	H	L	H	H
L	H	H	H	H	L	H	L	H	H
H	H	H	H	L	H	H	H	H	H

H = high level, L = low level

description

These full adders perform the addition of two 2-bit binary numbers. The sum (Σ) outputs are provided for each bit and the resultant carry (C2) is obtained from the second bit. Designed for medium-to-high-speed, multiple-bit, parallel-add/serial-carry applications, these circuits utilize high-speed, high-fan-out transistor-transistor logic (TTL) and are compatible with both DTL and TTL logic families. The implementation of a single-inversion, high-speed, Darlington-connected serial-carry circuit within each bit minimizes the necessity for extensive "look-ahead" and carry-cascading circuits.

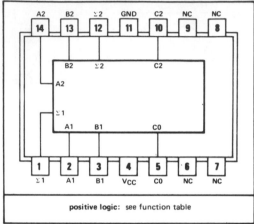

J OR N DUAL-IN-LINE OR W FLAT PACKAGE (TOP VIEW)

positive logic: see function table

NC—No internal connection

functional block diagram

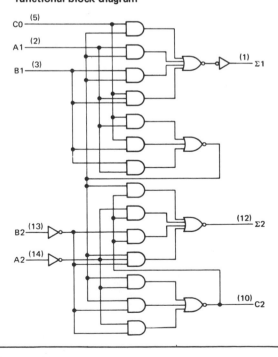

FIG. 11-11 TYPES SN5482, SN7482 2-BIT BINARY FULL ADDERS

- **For applications in:**
  Digital Computer Systems
  Data-Handling Systems
  Control Systems

- **SN54283/SN74283 Are Recommended For New Designs as They Feature Supply Voltage and Ground on Corner Pins to Simplify Board Layout**

TYPE	TYPICAL ADD TIMES		TYPICAL POWER DISSIPATION PER 4-BIT ADDER
	TWO 8-BIT WORDS	TWO 16-BIT WORDS	
'83A	23 ns	43 ns	310 mW
'LS83	89 ns	165 ns	75 mW

description

These full adders perform the addition of two 4-bit binary numbers. The sum (Σ) outputs are provided for each bit and the resultant carry (C4) is obtained from the fourth bit. The adders are designed so that logic levels of the input and output, including the carry, are in their true form. Thus the end-around carry is accomplished without the need for level inversion. Designed for medium-to-high-speed, the circuits utilize high-speed, high-fan-out transistor-transistor logic (TTL) but are compatible with both DTL and TTL families.

The '83A circuits feature full look ahead across four bits to generate the carry term in typically 10 nanoseconds to achieve partial look-ahead performance with the economy of ripple carry.

The 'LS83 can reduce power requirements to less than 20 mW/bit for power-sensitive applications. These circuits are implemented with single-inversion, high-speed, Darlington-connected serial-carry circuits within each bit.

Series 54 and 54LS circuits are characterized for operation over the full military temperature range of -55°C to 125°C; Series 74 and 74LS are characterized for 0°C to 70°C operation.

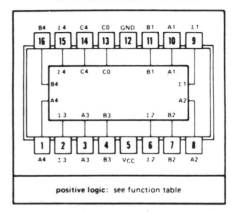

J OR N DUAL-IN-LINE OR W FLAT PACKAGE (TOP VIEW)

positive logic: see function table

FUNCTION TABLE

INPUT				OUTPUT					
				WHEN C0 = L			WHEN C0 = H		
						WHEN C2 = L			WHEN C2 = H
A1 / A3	B1 / B3	A2 / A4	B2 / B4	Σ1 / Σ3	Σ2 / Σ4	C2	Σ1 / Σ3	Σ2 / Σ4	C4
L	L	L	L	L	L	L	H	H	L
H	L	L	L	H	L	L	L	H	L
L	H	L	L	H	L	L	L	H	L
H	H	L	L	L	H	L	H	H	L
L	L	H	L	L	H	L	H	H	L
H	L	H	L	H	H	L	L	L	H
L	H	H	L	H	H	L	L	L	H
H	H	H	L	L	L	H	H	L	H
L	L	L	H	L	H	L	H	H	L
H	L	L	H	H	H	L	L	L	H
L	H	L	H	H	H	L	L	L	H
H	H	L	H	L	L	H	H	L	H
L	L	H	H	L	L	H	H	L	H
H	L	H	H	H	L	H	L	H	H
L	H	H	H	H	L	H	L	H	H
H	H	H	H	L	H	H	H	H	H

H = high level, L = low level

NOTE: Input conditions at A3, A2, B2, and C0 are used to determine outputs Σ1 and Σ2 and the value of the internal carry C2. The values at C2, A3, B3, A4, and B4 are then used to determine outputs Σ3, Σ4, and C4.

absolute maximum ratings over operating free-air temperature range (unless otherwise noted)

Supply voltage, V_{CC} (see Note 1)	7 V
Input voltage	5.5 V
Interemitter voltage (see Note 2)	5.5 V
Operating free-air temperature range: SN54', SN54LS' Circuits	-55°C to 125°C
SN74', SN74LS' Circuits	0°C to 70°C
Storage temperature range	-65°C to 150°C

NOTES: 1. Voltage values, except interemitter voltage, are with respect to network ground terminal.
2. This is the voltage between two emitters of a multiple-emitter transistor. For the '83A, this rating applies between the following pairs: A1 and B1, A2 and B2, A3 and B3, A4 and B4. For the 'LS83, this rating applies between the following pairs: A1 and B1, A1 and C0, B1 and C0, A3 and B3.

FIG. 11-12 TYPES SN5483A, SN54LS83, SN7483A, SN74LS83 4-BIT BINARY FULL ADDERS (Cont.)

functional block diagrams

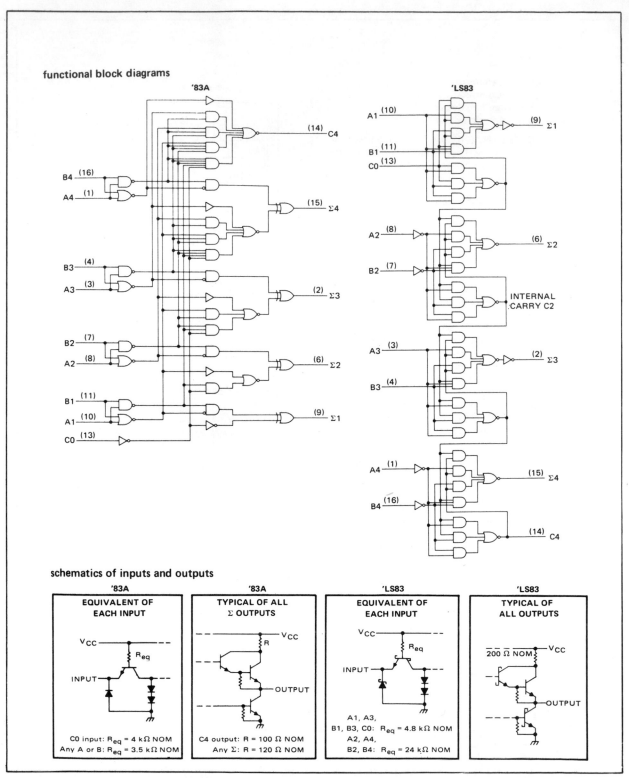

FIG. 11-12 TYPES SN5483A, SN54LS83, SN7483A, SN74LS83 4-BIT BINARY FULL ADDERS

complexity of these devices increases with their increasing ability to add more and more bits, each device uses the same basic logic concepts used by the simple adder and the half adder.

LABORATORY EXERCISE 11-1: INVESTIGATION OF HALF ADDER AND FULL ADDER CIRCUITS

PURPOSE

- To insure that the student understands the application of basic logic concepts and the logic operation of the adder circuits.

PROCEDURE

A. Construct the half adder circuit shown in figure 11-2.
B. To construct this circuit using NAND gates (such as type SN7400), use figure 11-13.
C. Verify the truth table for a half adder circuit.
D. Construct the full adder circuit of figure 11-8 using a type SN7480 IC. (Figure 11-14, page 98, shows the use of a NAND gate as an inverter.)
E. Verify the truth table for a full adder circuit.

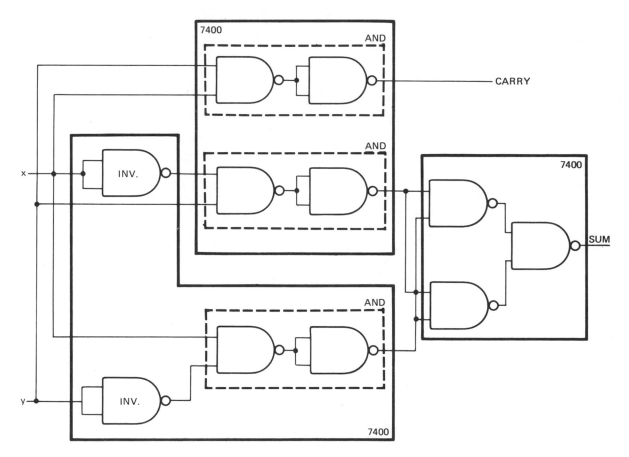

FIG. 11-13 THE HALF ADDER USING THREE NAND GATE ICS

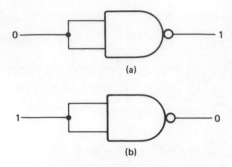

FIG. 11-14 NAND GATE AS INVERTER

EXTENDED STUDY TOPICS

1. Figure 11-13 shows the half adder circuit using NAND gates. Since a full adder circuit is simply two half adders connected in series,
    (a) draw the circuit (logic gate) of a full adder using NAND gates,
    (b) construct this circuit,
    (c) verify the truth table of this circuit.

2. Using one of the IC full adder devices described in this unit (do not use the 7480),
    (a) construct a full adder,
    (b) verify the truth table.

3. Using one of the IC full adders described in this unit,
    (a) draw the logic required if it is to be used as a subtracter,
    (b) construct the circuit,
    (c) verify the truth table for the circuit.

ANSWERS TO STUDENT REVIEW QUESTIONS

Unit 1 Generation of a Square Wave

(R1-1)   The pulse parameters depend upon the input conditions.

(R1-2)   The square wave has a 50% duty cycle; the rectangular wave does not.

(R1-3)   a. prt = 1 millisecond

           b. prr = 1000 Hz.

           c. 5 volts; actually, the value of $E_{avg.}$ is somewhat lower than 5 volts because the rise and fall times are not zero

(R1-4)   No, since the duty cycle is the relationship between the pulse width and the total pulse repetition time.

(R1-5)   Simplicity

Unit 2 Integrators and Differentiators

(R2-1) E_C = 12.9 volts

(R2-2) E_C = 13.95 volts

(R2-3) E_C = 2.09 volts

(R2-4) $t = 1.73 \times 10^{-4}$ second

(R2-5) $RC = 4.7 \times 10^{-3}$ second

(R2-6) $E_{C(2TC)}$ = 12.9 volts

(R2-7) $E_C(3 \times 10^{-3}$ seconds) = 7.09 volts

(R2-8) E_{cf} = 4.62 volts

(R2-9) E_C = 4.91 volts

(R2-10) $t_p \cong 10\ \tau$

(R2-11)   R and C and their tolerances

Unit 3 Clippers and Clampers

(R3-1) $R_L = 100\ k\Omega$

(R3-2) $E_{R_L} = 3.000$ volts

(R3-3)   The difference in the circuits is in the component across which the output is taken

(R3-4) R_S must be small compared to the load resistance

(R3-5)   The diode must be reversed

(R3-6) $R_d = 100\ k\Omega$

(R3-7) $C = 0.045 \mu F$

(R3-8) $\tau = 0.5$ sec.

(R3-9)   Positive clamping

(R3-10)   Yes

(R3-11)   The clipper circuit distorts the output waveform, while the clamper circuit does not distort the waveform

(R3-12)   Inadequate clamping action

Unit 5 The UJT Oscillator

(R5-1) $V_{R_{B_1}}$ = 7.68 V; $V_{R_{B_2}}$ = 4.32 V

(R5-2)   Silicon

Unit 6 The Schmitt Trigger

(R6-1) $t_p = \dfrac{1}{2000} \div 2 = 0.25$ millisecond

(R6-2) Voltage level (amplitude) detector

(R6-3) A typical switching transistor has a current gain of about 50; therefore, the contribution of base current to the emitter current is only about 2%.

(R6-4) $R_E = 250\Omega$

Unit 7 Multivibrators

(R7-1) This value will determine the value of I_B (or V_{BE}) needed to turn on each transistor. The minimum hfe must be known to determine the charge and discharge times, and therefore, t_{on} and t_{off} of the astable multivibrator.

(R7-2) $t_{pos} = 0.69\ RC = (0.69)(10 \times 10^3)(0.03 \times 10^{-6}) = 207\ \mu$ sec.

(R7-3) The duration of the pulse increases by a factor of four.

(R7-4) The monostable multivibrator is used in delay generation, pulse stretching, pulse synchronization, switch noise masking, etc.

(R7-5) Due to the fact that it appears to delay the pulse that causes it to function.

Unit 8 The Binary Number System

(R8-1) $111\ (7_{10})$

(R8-2) $1011\ (11_{10})$

(R8-3) $1100\ (12_{10})$

(R8-4) $010\ (2_{10})$

(R8-5) $100\ (4_{10})$

(R8-6) $001\ (1_{10})$

(R8-7) $64 + 4 + 2 = 70_{10}$

(R8-8) 1000100

(R8-9) (2) 0010
(5) 0101
(8) 1000
(9) 1001
(10) 1010

Unit 9 Boolean Algebra

(R9-1) $\overline{Y} + X$

(R9-2) $\overline{YZ} + X\overline{Z}$

(R9-3) $X\overline{Z} + \overline{YZ} + W\overline{Z}$

Unit 10 Logic Gates

(R10-1) No, but it is important to know the rules concerning the application of the logic family; all of the members of the logic family may be interconnected according to these rules.

(R10-2) Never leave a logic input disconnected. It makes a circuit noise sensitive.

(R10-3) All of the inputs of a NAND gate are connected together to obtain an inverter circuit.

(R10-4)

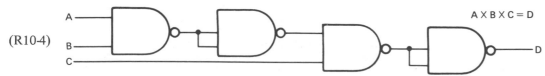

(R10-5) All of the pulses in figure 10-29 are examples of positive logic since the logic level is not concerned with voltage levels but rather with the assigning of zero and one logic levels.

Unit 11 Arithmetic Logic Gates

(R11-1)

A	0	1	0	1
B	0	0	1	1
C	0	1	1	0

(R11-2) Yes

(R11-3) A, X   1
         B, Y   0
         S   1
         C   0

(R11-4) Yes

(R11-5) Yes, since it cannot add a column of binary numbers greater than two digits.

(R11-6) The half adder does not have a provision for a carry input from a preceding stage while the full adder does have such a provision.

(R11-7) The half adder circuit cannot give a sum and carry out for all possible input conditions.

Acknowledgments

Contributions by Delmar Staff

Publications Director — Alan N. Knofla

Source Editor — Marjorie A. Bruce

Consulting Editor, Electronic Technology Series — Richard L. Castellucis

Technical Review — Jeffrey B. Duncan

Director of Manufacturing/Production — Frederick Sharer

Production Specialists — Patti Manuli, Debbie Monty, Sharon Lynch, Betty Michelfelder, Jean LeMorta, Alice Schielke, Lee St. Onge

Illustrators — Tony Canabush, George Dowse, Michael Kokernak, Al DeBenedetto

Appreciation is expressed to Texas Instruments Incorporated for its permission to reprint selected specifications.

Index